Immanuel Velikovsky was born in Vitebsk, Russia, in 1895 and has studied at the universities of Moscow, Berlin, Vienna and Edinburgh. From 1921 to 1924 he edited, with Albert Einstein, the Scripta Universitatis atque Bibliothecae Hierosolymitarum, from which the Hebrew University of Jerusalem was to grow. In 1939 he emigrated to America where he still lives, in Princeton, New Jersey.

His first controversial book, *Worlds in Collision*, was published in 1950 and was followed by *Ages in Chaos* (Vol 1, 1952), *Earth in Upheaval* (1955) and *Oedipus and Akhnaton* (1960).

Also by Immanuel Velikovsky and available in Abacus

WORLDS IN COLLISION
AGES IN CHAOS

Immanuel Velikovsky

EARTH IN
UPHEAVAL

ABACUS edition published in 1973
by Sphere Books Ltd
30/32 Gray's Inn Road, London WC1X 8JL
Reprinted 1974, 1976
First published in Great Britain
by Victor Gollancz Ltd in association with
Sidgwick & Jackson Ltd 1956
Copyright © 1955 by Immanuel Velikovsky

To my daughters
Shulamith and Ruth

Set in Linotype Times Roman

*Printed in Great Britain by Hazell Watson & Viney Ltd
Aylesbury, Bucks*

PREFACE

Earth in Upheaval is a book about the great tribulations to which the planet on which we travel was subjected in pre-historical and historical times. The pages of this book are transcripts of the testimony of mute witnesses, the rocks, in the court of celestial traffic. They testify by their own appearance and by the encased contents of dead bodies, fossilized skeletons. Myriads upon myriads of living creatures came to life on this ball of rock suspended in nothing and returned to dust. Many died a natural death, many were killed in wars between races and species, and many were entombed alive during great paroxysms of nature in which land and sea contested in destruction. Whole tribes of fish that had filled the oceans suddenly ceased to exist; of entire species and even genera of land animals not a single survivor was left.

The earth and the water without which we cannot exist suddenly turned into enemies and engulfed the animal kingdom, the human race included, and there was no shelter and no refuge. In such cataclysms the land and the sea repeatedly changed places, laying dry the kingdom of the ocean and submerging the kingdoms of the land.

In *Worlds in Collision* I presented the chronicles of two—the very last—series of such catastrophes, those that visited our earth in the second and first millennia before the present era. Since these upheavals occurred in historical times, when the art of writing had already been perfected in the centres of ancient civilization, I described them mainly from historical documents, relying on celestial charts, calendars, and sundials and water clocks discovered by archaeologists, and drawing also upon classical literature, the sacred literature of East and West, the epics of the northern races, and the oral traditions of primitive peoples from Lapland to the South Seas. Geological vestiges of the events narrated in documents and traditions were indicated only here and there, when I felt that the immediate testimony of the rocks must be presented along with the historical evidence. I closed that description of cataclysmic events with a promise to attempt, at a later date, the reconstruction of similar global catastrophes of earlier times, one of them being the Deluge.

I had intended, after piecing together the history of these earlier global upheavals, to present geological and paleontological material to support the testimony of man. But the reception of *Worlds in Collision* by certain scientific groups persuaded me, before reviving the pageant of earlier catastrophes, to present at least some of the evidence of the rocks, which is as insistent as that carried down to our times by written records and by word of mouth. This testimony is never given in metaphors; and as with the pages of the Old Testament or of the *Iliad*, nothing can be changed in it. Pebbles and rocks and mountains and the bottom of the sea will bear witness. Do they know of the days, recent and ancient, when the harmony of this world was interrupted by the forces of nature? Have they entombed innumerable creatures and encased them in rock? Have they seen the ocean moving on continents and continents sliding under water? Was this earth and the expanse of its seas showered with stones and covered by ashes? Were its forests, uprooted by hurricanes and set afire, covered by tides carrying sand and debris from the bottom of the oceans? It takes millions of years for a log to be turned into coal but only a single hour when burning. Here lies the core of the problem: Did the earth change in a slow process, a year added to a year and a million years to a million, the peaceful ground of nature being the broad arena of the contest of throngs, in which the fittest survived? Or did it happen, too, that the very arena itself, infuriated, rose against the contestants and made an end of their battles?

I present here some pages from the book of nature. I have excluded from them all references to ancient literature, traditions, and folklore; and this I have done with intent, so that careless critics cannot decry the entire work as "tales and legends." Stones and bones are the only witnesses. Mute as they are, they will testify clearly and unequivocally. Yet dull ears and dimmed eyes will deny this evidence, and the dimmer the vision, the louder and more insistent will be the voices of protestation. This book was not written for those who swear by the *verba magistri*—the holiness of their school wisdom; and they may debate it without reading it, as well.

ACKNOWLEDGMENTS

Working on *Earth in Upheaval* and on the essay (Address before the Graduate College Forum of Princeton University) added at the end of this volume, I have incurred a debt of gratitude to several scientists.

Professor Walter S. Adams, for many years director of Mount Wilson Observatory, gave me all the information and instruction for which I asked concerning the atmospheres of the planets, a field in which he is the outstanding authority. On my visit to the solar observatory in Pasadena, California, and in our correspondence he has shown a fine spirit of scientific cooperation.

The late Dr. Albert Einstein, during the last eighteen months of his life (November 1953–April 1955), gave me much of his time and thought. He read several of my manuscripts and supplied them with marginal notes. Of *Earth in Upheaval* he read chapters VIII to XII; he made handwritten comments on this and other manuscripts and spent not a few long afternoons and evenings, often till midnight, discussing and debating with me the implications of my theories. In the last weeks of his life he reread *Worlds in Collision* and read also three files of "memoirs" on that book and its reception, and expressed his thoughts in writing. We started at opposite points; the area of disagreement, as reflected in our correspondence, grew ever smaller, and though at his death (our last meeting was nine days before his passing) there remained clearly defined points of disagreement, his stand then demonstrated the evolution of his opinion in the space of eighteen months.

Professor Waldo S. Glock, Chairman of the Department of

Geology at Macalester College, St. Paul, Minnesota, a recognized authority in dendrochronology (dating of tree rings), with the help of his graduate students searched the literature pertaining to the tree rings of early ages, and also gave me answers to questions in his field.

Dr. H. Manley of the Imperial College, London, Professor P. L. Mercanton of the University of Lausanne, and Professor E. Thellier of the Observatoire Géophysique of the University of Paris, gave me freely of their knowledge in the field of geomagnetism and sent me reprints of their works.

Professor Lloyd Motz of the Department of Astronomy at Columbia University, New York, never tired of testing mathematically and of commenting on various problems in electromagnetism and in celestial mechanics which I offered for discussion.

Dr. T. E. Nikulins, geologist in Caracas, Venezuela, repeatedly drew my attention to various publications in the scientific press that might be of help to me; he supplied me with the source dealing with the discovery of the stone and bronze ages in northeastern Siberia.

Professor George McCready Price, geologist in California, read an early draft of various chapters of this work. Between this octogenarian, author of several books on geology written from the fundamentalist point of view, and myself, there are some points of agreement and as many of disagreement. The main one among the latter is that while Price is opposed to the very theory of evolution and is supported in his disbelief by the fact that since the scientific age no new animal species have been observed to emerge, I offer in the concluding chapters of this book ("Extinction" and "Cataclysmic Evolution") a radical solution of the problem.

With Professor Richardson of the Illinois Institute of Technology I spent several days discussing a few problems in physics and geophysics.

With no one do I share the responsibility for my work; to everyone who gave me a helpful hand while the atmosphere in academic circles was generally charged with animosity, I express here my gratitude.

CONTENTS

Earth in Upheaval

Chapter I

IN THE NORTH

In Alaska

In Alaska, to the north of Mount McKinley, the tallest mountain in North America, the Tanana River joins the Yukon. From the Tanana Valley and the valleys of its tributaries gold is mined out of gravel and "muck." This muck is a frozen mass of animals and trees.

F. Rainey of the University of Alaska described the scene[1]: "Wide cuts, often several miles in length and sometimes as much as 140 feet in depth, are now being sluiced out along stream valleys tributary to the Tanana in the Fairbanks District. In order to reach gold-bearing gravel beds an over-burden of frozen silt or 'muck' is removed with hydraulic giants. This 'muck' contains enormous numbers of frozen bones of extinct animals such as the mammoth, mastodon, super-bison and horse."[2]

These animals perished in rather recent times; present estimates place their extinction at the end of the Ice Age or in early post-glacial times. The soil of Alaska covered their bodies together with those of animals of species still surviving.

Under what conditions did this great slaughter take place, in which millions upon millions of animals were torn limb from limb and mingled with uprooted trees?

F. C. Hibben of the University of New Mexico writes: "Although the formation of the deposits of muck is not clear,

[1] F. Rainey, "Archaeological Investigation in Central Alaska," *American Antiquity*, V (1940), 305.
[2] The horse became extinct in pre-Columbian America; the present horses in the Western Hemisphere are descendants of imported animals.

there is ample evidence that at least portions of this material were deposited under catastrophic conditions. Mammal remains are for the most part dismembered and disarticulated, even though some fragments yet retain, in their frozen state, portions of ligaments, skin, hair, and flesh. Twisted and torn trees are piled in splintered masses. . . . At least four considerable layers of volcanic ash may be traced in these deposits, although they are extremely warped and distorted. . . ."[3]

Could it be that a volcanic eruption killed the animal population of Alaska, the streams carrying down into the valleys the bodies of the slaughtered animals? A volcanic eruption would have charred the trees but would not have uprooted and splintered them; if it killed animals, it would not have dismembered them. The presence of volcanic ash indicates that a volcanic eruption did take place, and repeatedly, in four consecutive stages of the same epoch; but it is also apparent that the trees could have been uprooted and splintered only by hurricane or flood or a combination of both agencies. The animals could have been dismembered only by a stupendous wave that lifted and carried and smashed and tore and buried millions of bodies and millions of trees. Also, the area of the catastrophe was much greater than the action of a few volcanoes could have covered.

Muck deposits like those of the Tanana River Valley are found in the lower reaches of the Yukon in the western part of the peninsula, on the Koyukuk River that flows into the Yukon from the north, on the Kuskokwim River that empties its waters into Bering Sea, and at several places along the Arctic coast, and so "may be considered to extend in greater or lesser thickness over all unglaciated areas of the northern peninsula."[4]

What could have caused the Arctic Sea and the Pacific Ocean to irrupt and wash away forests with all their animal population and throw the entire mingled mass in great heaps scattered all over Alaska, the coast of which is longer than the Atlantic sea-board from Newfoundland to Florida? Was it not a tectonic revolution in the earth's crust, that also caused the

[3] F. C. Hibben, "Evidence of Early Man in Alaska," *American Antiquity*, VIII (1943), 256.
[4] Ibid.

4

volcanoes to erupt and to cover the peninsula with ashes? In various levels of the muck, stone artifacts were found "frozen *in situ* at great depths and in apparent association" with the Ice Age fauna, which implies that "men were contemporary with extinct animals in Alaska."[5] Worked flints, characteristically shaped, called Yuma points, were repeatedly found in the Alaskan muck, one hundred and more feet below the surface. One such spear point was found there between a lion's jaw and a mammoth's tusk.[6] Similar weapons were used only a few generations ago by the Indians of the Athapascan tribe, who camped in the upper Tanana Valley.[7] "It has also been suggested that even modern Eskimo points are remarkably Yuma-like,"[8] all of which indicates that the multitudes of torn animals and splintered forests date from a time not many thousand years ago.

The Ivory Islands

The arctic coast of Siberia is cold, bleak, inhospitable. The sea is passable for ships manoeuvring between floating ice for two months of the year; from September to the middle of July the ocean north of Siberia is fettered, an unbroken desert of ice. Polar winds sweep over the frozen tundras of Siberia, where no tree grows and the soil is never tilled. In his exploratory voyage on the ship *Vega* in 1878, Nils Adolf Erik Nordenskjöld, the first to traverse this northern seaway from one end to the other, travelled for weeks along the coast from Novaya Zemlya to Cape Shelagskoi (170° 30′ East) on the eastern extremity of Siberia without seeing a single human being on the shore.

Fossil tusks of the mammoth—an extinct elephant—were found in northern Siberia and brought southward to markets at a very early time, possibly in the days of Pliny in the first century of the present era. The Chinese excelled in working delicate designs in the ivory, much of which they obtained from the north. And from the days of the conquest of Siberia (1582) by the Cossack Yermak under Ivan the Terrible, until our own

[5] Rainey, *American Antiquity*, V. 307.
[6] Hibben, *American Antiquity*, VIII, 257.
[7] Rainey, *American Antiquity*, V, 301.
[8] Hibben, *American Antiquity*, VIII, 256.

5

times, trade in mammoths' tusks has gone on. Northern Siberia provided more than half the world's supply of ivory, many piano keys and many billiard balls being made from the fossil tusks of these mammoths.

In 1797, the body of a mammoth, with flesh, skin, and hair, was found in northeastern Siberia, and since then bodies of other mammoths have been unearthed from the frozen ground in various parts of that region. The flesh had the appearance of freshly frozen beef; it was edible, and wolves and sledge dogs fed on it without harm.[1]

The ground must have been frozen ever since the day of their entombment; had it not been frozen, the bodies of the mammoths would have putrified in a single summer, but they remained unspoiled for some thousands of years. "It is therefore absolutely necessary to believe that the bodies were frozen up immediately after the animals died, and *were never once thawed*, until the day of their discovery."[2]

High in the north above Siberia, six hundred miles inside the Polar Circle, in the Arctic Ocean, lie the Liakhov Islands. Liakhov was a hunter who, in the days of Catherine II, ventured to these islands and brought back the report that they abounded in mammoths' bones. "Such was the enormous quantity of mammoths' remains that it seemed . . . that the island was actually composed of the bones and tusks of elephants, cemented together by icy sand."[3]

The New Siberian Islands, discovered in 1805 and 1806, as well as the islands of Stolbovoi and Belkov to the west, present the same picture. "The soil of these desolate islands is absolutely packed full of the bones of elephants and rhinoceroses in astonishing numbers."[4] "These islands were full of mammoth bones, and the quantity of tusks and teeth of elephants and rhinoceroses, found in the newly discovered island of New Siberia, was perfectly amazing, and surpassed anything which had as yet been discovered."[5]

Did the animals come there over the ice, and for what purpose? On what food could they have lived? Not on the lichens

[1] Observation of D. F. Hertz, in B. Digby, *The Mammoth* (1926), p. 9.
[2] D. Gath Whitley, "The Ivory Islands in the Arctic Ocean," *Journal of the Philosophical Society of Great Britain*, XII (1910), 35.
[3] Ibid., p. 41. [4] Ibid., p. 36. [5] Ibid., p. 42.

of the Siberian tundras, covered by deep snow most of the year, and still less on the moss of the polar islands, which are frozen ten months in the year: mammoths, members of the voracious elephant family, required huge quantities of vegetable food every day in the year. How could large herds of them have existed in a country like northeast Siberia, which is regarded as the coldest place in the world, and where there was no food for them?

Mammoth tusks have been dredged in nets from the bottom of the Arctic Ocean; and after arctic gales the shores of the islands are strewn with tusks cast up by the billows. This is regarded as an indication that the bottom of the Arctic Ocean between the islands and the mainland was dry land in the days when mammoths roamed there.

Georges Cuvier, the great French paleontologist (1769–1832), thought that in a vast catastrophe of continental dimensions the sea overwhelmed the land, the herds of mammoths perished, and in a second spasmodic movement the sea rushed away, leaving the carcasses behind. This catastrophe must have been accompanied by a precipitous drop in temperature; the frost seized the dead bodies and saved them from decomposition.[6] In some mammoths, when discovered, even the eyeballs were still preserved.

Charles Darwin, who denied the occurrence of continental catastrophes in the past, in a letter to Sir Henry Howorth admitted that the extinction of mammoths in Siberia was for him an insoluble problem.[7] J. D. Dana, the leading American geologist of the second half of the last century, wrote: "The encasing in ice of huge elephants, and the perfect preservation of the flesh, shows that the cold finally became *suddenly* extreme, as of a single winter's night, and knew no relenting afterward."[8]

In the stomachs and between the teeth of the mammoths were found plants and grasses that do not grow now in northern Siberia. "The contents of the stomachs have been carefully examined; they showed the undigested food, leaves of trees now found in Southern Siberia, but a long way from the existing

[6] Georges Cuvier, *Discours sur les révolutions de la surface du globe et sur les changements qu'elles ont produits dans le règne animal* (1825).

[7] Whitley, *Journal of the Philosophical Society of Great Britain*, XII (1910), 56. G. F. Kunz, *Ivory and the Elephant* (1916), p. 236.

[8] J. D. Dana, *Manual of Geology* (4th ed.; 1894), p. 1007.

deposits of ivory. Microscopic examination of the skin showed red blood corpuscles, which was a proof not only of a sudden death, but that the death was due to suffocation either by gases or water, evidently the latter in this case. But the puzzle remained to account for the sudden freezing up of this large mass of flesh so as to preserve it for future ages."[9]

What could have caused a sudden change in the temperature of the region? Today the country does not provide food for large quadrupeds, the soil is barren and produces only moss and fungi a few months in the year; at that time the animals fed on plants. And not only mammoths pastured in northern Siberia and on the islands of the Arctic Ocean. On Kotelnoi Island "neither trees, nor shrubs, no bushes, exist . . . and yet the bones of elephants, rhinoceroses, buffaloes, and horses are found in this icy wilderness in numbers which defy all calculation."[10]

When Hedenström and Sannikov discovered the New Siberian Islands in 1806, they found in the "desolate wilderness" of polar sea the remains of "enormous petrified forests." These forests could be seen tens of miles away. "The trunks of the trees in these ruins of ancient forests were partly standing upright and partly lying horizontally buried in the frozen soil. Their extent was very great."[11] Hedenström described them as follows: "On the southern coast of New Siberia are found the remarkable wood hills [piles of trunks]. They are 30 fathoms [180 feet] high, and consist of horizontal strata of sandstone, alternating with strata of bituminous beams or trunks of trees. On ascending these hills, fossilized charcoal is everywhere met with, covered apparently with ashes; but, on closer examination, this ash is also found to be a petrifaction, and so hard that it can scarcely be scraped off with a knife."[12] Some trunks are fixed perpendicularly in the sandstone, with broken ends.

In 1829 the German scientist G. A. Erman went to the Laikhov and the New Siberian Islands to measure there the magnetic field of the earth. He described the soil as full of the bones of elephants, rhinoceroses, and buffaloes. Of the piles of wood he

[9] Whitley, *Journal of the Philosophical Society of Great Britain*, XII (1910), 56.
[10] Ibid., p. 50. [11] Ibid., p. 43.
[12] F. P. Wrangell, *Narrative of an Expedition to Siberia and the Polar Sea*, (1841), note to p. 173 of the American edition.

wrote: "In New Siberia [Island], on the declivities facing the south, lie hills 250 or 300 feet high, formed of driftwood, the ancient origin of which, as well as of the fossil wood in the tundras, anterior to the history of the Earth in its present state, strikes at once even the most uneducated hunters. . . . Other hills on the same island, and on Kotelnoi, which lies further to the west, are heaped up to an equal height with skeletons of pachyderms [elephants, rhinoceroses], bisons, etc., which are cemented together by frozen sand as well as by strata and veins of ice. . . . On the summit of the hills they [the trunks of trees] lie flung upon one another in the wildest disorder, forced upright in spite of gravitation, and with their tops broken off or crushed, as if they had been thrown with great violence from the south on a bank, and there heaped up."[13]

Eduard von Toll repeatedly visited the New Siberian Islands from 1885 to 1902, when he perished in the Arctic Ocean. He examined the "wood hills" and "found them to consist of carbonized trunks of trees, with impressions of leaves and fruits."[14] On Maloi, one of the group of Liakhov Islands, Toll found bones of mammoths and other animals together with the trunks of fossil trees, with leaves and cones. "This striking discovery proves that in the days when the mammoths and rhinoceroses lived in northern Siberia, these desolate islands were covered with great forests, and bore a luxuriant vegetation."[15]

A hurricane, apparently, uprooted the trees of Siberia and flung them to the extreme north; mountainous waves of the ocean piled them in huge hills, and some agent of a bituminous nature transformed them into charcoal, either before or after they were deposited and cemented in drifted masses of sand that became baked into sandstone.

These petrified forests were swept from northern Siberia into the ocean, and together with bones of animals and drifted sand built the islands. It may be that not all the charred trees and the mammoths and other animals were destroyed and swept away in a single catastrophe. It is more probable that one huge cemetery of animals and trees came flying through the air on the

[13] G. A. Erman, *Travels in Siberia* (1848), II, 376, 383.
[14] Whitley, *Journal of the Philosophical Society of Great Britain*, XII (1910), 49.
[15] Ibid., p. 50.

crest of a retreating tidal wave to settle astride another, older, cemetery, deep in the Polar Circle.

The scientists who explored the "muck" beds of Alaska have not reflected upon the similarity in appearance of animal remains there and in the polar regions of Siberia and on arctic islands, and have therefore not discussed a common cause. The exploration of the New Siberian Islands, one thousand miles away from Alaska, was the work of eighteenth- and nineteenth-century academicians who followed the hunters of fossil ivory; the exploration of Alaskan soil was the work of twentieth-century scientists who followed the gold-digging machines.

These two observations—one old, one new—came from the north. Before presenting many more from all parts of the world, I shall review a few dominant theories on the history of our earth and its animal kingdom. We shall read in brief, in the original statements of the authors, how the earlier naturalists explained the phenomena; how, subsequently, the same phenomena were interpreted in terms of slow evolution; and how in the last fourscore years more and more facts have presented themselves that do not square with the picture of a peaceful world moulded in a slow and uneventful process.

Chapter II

REVOLUTION

The Erratic Boulders

"The waters of the ocean in which our mountains had been formed still covered a part of these Alps when a violent paroxysm of the globe suddenly opened great cavities . . . and ruptured many rocks. . . .

"The waters were carried toward these abysses with extreme violence, falling from the height they were before; they crossed deep valleys and dragged immense quantities of earth, sand, and debris of all kinds of rocks. This mass, shoved along by the onrush of great waters, was left spread up the slopes where we still see many scattered fragments."[1]

Thus did Horace Bénédict de Saussure, foremost Swiss naturalist of the end of the eighteenth century, explain the presence of stones broken off from the Alps and carried to the Jura Mountains to the west; so also did he explain the marine remains found in alpine ridges, and the sand, gravel, and clay that fill the valleys of the Alps and the plains beyond them.

The loose rocks lying on the Jura Mountains were torn from the Alps; in their mineral composition they differ from the rock formations of the Jura, showing their alpine origin. Rocks that differ from the formations on which they lie are called "erratic boulders."

These stone blocks lie on the Jura Mountains at an elevation of 2000 feet above Lake Geneva. Some of them are thousands of cubic feet in size, and Pierre à Martin is over 10,000 cubic feet. They must have been carried across the space now occupied by the lake and lifted to the height where they are found.

[1] Horace Bénédict de Saussure, *Voyages dans les Alpes*, I (1779), 151.

There are erratic boulders in many places of the world. In the British Isles, on the shore and in the highlands, are enormous quantities of them, transported there across the North Sea from the mountains of Norway. Some force wrested them from those massifs, bore them over the entire expanse that separates Scandinavia from the British Isles, and set them down on the coast and on the hills. From Scandinavia boulders were also carried to Germany and spread over that country, in some places so thickly that it seems as though they had been brought there by masons to build cities. Also, high in the Harz Mountains, in central Germany, lie stones that originated in Norway.

From Finland blocks of stone were swept to the Baltic regions and over Poland and lifted onto the Carpathians. Another train of boulders was fanned out from Finland, over the Valdai Hills, over the site of Moscow, and as far as the Don.

In North America erratic blocks, broken from the granite of Canada and Labrador, were spread over Maine, New Hampshire, Vermont, Massachusetts, Connecticut, New York, New Jersey, Michigan, Wisconsin, and Ohio; they perch on top of ridges and lie on slopes and deep in the valleys. They lie on the coastal plain and on the White Mountains and the Berkshires, sometimes in an unbroken chain; in the Pocono Mountains they balance precariously on the edge of crests. The attentive traveller through the woods wonders at the size of these rocks, brought there and abandoned sometime in the past, frighteningly piled up.

Some erratics are enormous. The block near Conway, New Hampshire, is 90 by 40 by 38 feet and weighs about 10,000 tons, the load of a large cargo ship. Equally large is Mohegan Rock, which towers over the town of Montville, in Connecticut. The great flat erratic in Warren County, Ohio, weighs approximately 13,500 tons and covers three quarters of an acre; the Ototoks erratic, thirty miles south of Calgary, Alberta, consists of two pieces of quartzite "derived from at least 50 miles to the west," of a calculated weight of over 18,000 tons.[2] Blocks of 250 to 300 feet in circumference, however, are small when compared with a mass of chalk stone near Malmö in southern Sweden, which is "three miles long, one thousand feet wide and from one hundred to two hundred feet in thickness, and which

[2] R. F. Flint, *Glacial Geology and the Pleistocene Epoch* (1947), pp. 116–17.

has been transported an indefinite distance. . . ." It is quarried for commercial purposes. A similar transported slab of chalk is found on the eastern coast of England, "upon which a village had unwittingly been built."[3]

In innumerable places on the surface of the earth, as well as on isolated islands in the Atlantic and Pacific and in Antarctica,[4] lie rocks of foreign origin, brought from afar by some great force. Broken off from their parent mountain ridges and coastal cliffs, they were carried down dale and up hill and over land and sea.

Sea and Land Changed Places

The most renowned naturalist to come from the generation of the French Revolution and the Napoleonic Wars was Georges Cuvier. He was the founder of vertebrate paleontology, or the science of fossil bones, and thus of the science of extinct animals. Studying the finds made in the gypsum formation of Montmartre in Paris and those elsewhere in France and the European continent in general, he came to the conclusion that in the midst of even the oldest strata of marine formations there are other strata replete with animal or plant remains of terrestrial or fresh-water forms; and that among the more recent strata, or those that are nearer the surface, there are also land animals buried under heaps of marine sediment. "It has frequently happened that lands which have been laid dry, have been again covered by the waters, in consequence either of their being engulfed in the abyss, or of the sea having merely risen over them. . . . These repeated irruptions and retreats of the sea have neither all been slow nor gradual; on the contrary, most of the catastrophes which have occasioned them have been sudden; and this is especially easy to be proven, with regard to the last of these catastrophes, that which, by a twofold motion, has inundated, and afterwards laid dry, our present continents, or at least a part of the land which forms them at the present day.[1]

"The breaking to pieces, the raising up and overturning of

[3] G. F. Wright, *The Ice Age in North America and Its Bearing upon the Antiquity of Man* (5th ed; 1911), pp. 238–39.
[4] E. H. Shackleton, *The Heart of the Antarctic*, II (1909), illustration opposite p. 293.
[1] Georges Cuvier, *Essay on the Theory of the Earth* (5th ed; 1827) (English translation of *Discours sur les révolutions de la surface du globe*), pp. 13–14.

the older strata [of the earth], leave no doubt upon the mind that they have been reduced to the state in which we now see them, by the action of sudden and violent causes; and even the force of the motions excited in the mass of waters, is still attested by the heaps of debris and rounded pebbles which are in many places interposed between the solid strata. Life, therefore, has often been disturbed on this earth by terrific events. Numberless living beings have been the victims of these catastrophes; some, which inhabited the dry land, have been swallowed up by inundations; others, which peopled the waters, have been laid dry, the bottom of the sea having been suddenly raised; their very races have been extinguished for ever, and have left no other memorial of their existence than some fragments which the naturalist can scarcely recognize."[2]

Cuvier was surprised to find that "life has not always existed upon the globe," for there are deep strata which contain no vestiges of living beings. The sea without inhabitants "would seem to have prepared materials for the mollusca and zoophytes," and when they appeared and populated the sea, they deposited their shells and built coral, at first in small numbers, and eventually in vast formations.

Cuvier believed that changes have operated in nature not just since the appearance of life, for the land masses formed previous to that event also seemed to have experienced violent displacements.[3]

He found in the gypsum deposits in the suburbs of Paris marine limestone containing over eight hundred species of shells, all of them marine. Under this limestone there is another —fresh-water—deposit formed of clay. Among the shells, all of fresh-water (or land) origin, there are also bones—but "what is remarkable," the bones are those of reptiles and not of mammals, "of crocodiles and tortoises."

Much of France was once sea; then it was land, populated by land reptiles; then it became sea again and was populated by marine animals; then it was land again, inhabited by mammals; then it was once more sea, and again land. Each stratum contains the evidence of its age in the bones and shells of the animals that lived and propagated there at the time and were entombed in recurrent upheavals. And as it was on the site of Paris, so it

[2] Ibid., p. 15. [3] Ibid., p. 20.

14

was in other parts of France, and in other countries of Europe.

The strata of the earth disclose that "The thread of operations is here broken; the march of Nature is changed; and none of the agents which she now employs, would have been sufficient for the production of her ancient works."[4]

"We have no evidence that the sea can now encrust those shells with a paste as compact as that of the marbles, the sandstones, or even the coarse limestone. . . .

"In short, all [now active] causes united, would not change, in an appreciable degree, the level of the sea; nor raise a single stratum above its surface. . . . It has been asserted that the sea has undergone a general diminishing of level. . . . Admitting that there has been a gradual diminution of the waters; that the sea has transported solid matter in all directions; that the temperature of the globe is either diminishing or increasing; none of these cases could have overturned our strata, enveloped in ice large animals, with their flesh and skin; laid dry marine [animals] . . . and, lastly, destroyed numerous species, and even entire genera."[5]

"Thus, we repeat, it is in vain that we search, among the powers which now act at the surface of the earth, for causes sufficient to produce the revolutions and catastrophes, the traces of which are exhibited by its crust."[6]

But what could have caused these catastrophes? Cuvier reviewed the theories of the origin of the world current in his time but found no answer to the question that preoccupied him. He did not know the cause of these vast cataclysms; he only knew that they had occurred. "Many fruitless efforts" had been made, and he felt that his search for the causes of the cataclysms was fruitless too. "These ideas have haunted, I may almost say have tormented me during my researches among fossil bones."[7]

The Caves of England

In 1823, William Buckland, professor of geology at the University of Oxford, published his *Reliquiae diluvianae* (*Relics of the*

[4] Ibid., p. 24.   [5] Ibid., pp. 32, 36–37.   [6] Ibid., pp. 35–36.
[7] Ibid., p. 242.

Flood), with the subtitle, *Observations on the organic remains contained in caves, fissures, and diluvial gravel, and on other geological phenomena, attesting the action of an universal deluge.* Buckland was one of the great authorities on geology of the first half of the nineteenth century. In a cave in Kirkdale in Yorkshire, eighty feet above the valley, under a floor covering of stalagmites, he found teeth and bones of elephants, rhinoceroses, hippopotami, horses, deer, tigers (the teeth of which were "larger than those of the largest lion or Bengal tiger"), bears, wolves, hyenas foxes, hares, rabbits, as well as bones of ravens, pigeons, larks, snipe, and ducks. Many of the animals had died "before the first set, or milk teeth, had been shed."

Certain scholars prior to Buckland had their own explanation for the provenience of elephant bones in the soil of England, and to them Buckland referred: "[The idea] which long prevailed, and was considered satisfactory by the antiquaries [archaeologists] of the last century, was, that they were the remains of elephants imported by the Roman armies. This idea is also refuted: First, by the anatomical fact of their belonging to an extinct species of this genus; second, by their being usually accompanied by the bones of rhinoceros and hippopotamus, animals which could never have been attached to Roman armies; thirdly, by their being found dispersed over Siberia and North America, in equal or even greater abundance than in those parts of Europe which were subjected to the Roman power."[1]

It appeared that hippopotamus and reindeer and bison lived side by side at Kirkdale; hippopotamus, reindeer, and mammoth pastured together at Brentford near London.[2] Reindeer and grizzly bear lived with the hippopotamus at Cefn in Wales. Lemming and reindeer bones were found together with bones of the cave lion and hyena at Bleadon in Somerset.[3] Hippopotamus, bison, and musk sheep were found together with worked flint in the gravels of the Thames Valley.[4] The remains of reindeer lay with the bones of mammoth and rhinoceros in the cave of Breugue in France, in the same red clay, encased by

[1] W. Buckland, *Reliquiae diluvianae*, p. 173.
[2] W. B. Dawkins, *Proceedings of the Geological Society* (1869), p. 190.
[3] Ibid.
[4] James Geikie, *Prehistoric Europe* (1881), p. 137; Dawkins, *Cave-hunting* (1874), p. 416.

the same stalagmites.[5] At Arcy, France, also in a cave, bones of the hippopotamus were found with bones of the reindeer, and with them a worked flint.[6]

According to the prophecy of Isaiah (11:6), in messianic times to come the lion and the calf would pasture together. But even prophetic vision has not conceived of a reindeer from snow-covered Lapland and a hippopotamus from the tropical Congo River living together on the British Isles or in France. Yet they did leave their bones in the same mud of the same caves, together with bones of other animals, in the strangest assortments.

These animal bones were found in gravel and clay to which Buckland gave the name of diluvium.

Buckland was concerned "to establish two important facts, first, that there has been a recent and general inundation of the globe; and, second, that the animals whose remains are found interred in the wreck of that inundation were natives of high north latitudes." The presence of tropical animals in northern Europe "cannot be solved by supposing them to migrate periodically . . . for in the case of crocodiles and tortoises extensive emigration is almost impossible, and not less so to such an unwieldy animal as the hippopotamus when out of the water." But how could they live in the cold of northern Europe? Buckland says: "It is equally difficult to imagine that they could have passed their winters in lakes or rivers frozen up with ice." If cold-blooded land animals are unable to hide themselves in the ground over the winter, in icy climates their blood would freeze solid: they lack the ability to regulate the temperature of their bodies. Like Cuvier, Buckland was "nearly certain that if any change of climate has taken place, it took place suddenly."[7]

Of the time the catastrophe occurred, which covered with mud and pebbles the bones in the Kirkdale cave, Buckland wrote: "From the limited quantity of postdiluvian stalactite, as well as from the *undecayed condition of the bones*," one must deduce that "the time elapsed since the introduction of the diluvial mud has not been of excessive length." The bones were not

[5] Cuvier, *Recherches sur les ossements fossiles des quadrupèdes*, IV, 94.
[6] E. Lartet, *Reliquiae aquitanicae*, pp. 147–48.
[7] Buckland, *Reliquiae diluvianae*, p. 47.

17

yet fossilized; their organic matter was not yet replaced by minerals. Buckland thought that the time elapsed since a diluvial catastrophe could not have exceeded five or six thousand years, the figure adopted also by De Luc, Dolomieu, and Cuvier, each of whom presented his own reasons.

Then the illustrious geologist added these words: "What [the] cause was, whether a change in the inclination in the earth's axis, or the near approach of a comet, or any other cause or combination of causes purely astronomical, is a question the discussion of which is foreign to the object of the present memoir."

The Aquatic Graveyards

The Old Red Sandstone is regarded as one of the oldest strata with signs of extinct life in it. No animal life higher than fish is found there. Whatever the age of this formation, it carries the testimony and "a wonderful record of violent death falling at once, not on a few individuals, but on whole tribes."[1]

In the late thirties of the last century Hugh Miller made the Old Red Sandstone in Scotland the special subject of his investigations. He observed: "The earth had already become a vast sepulchre, to a depth beneath the bed of the sea equal to at least twice the height of Ben Nevis over its surface."[2] Ben Nevis in the Grampian Mountains is the highest peak in Great Britain, 4406 feet high. The stratum of the Old Red Sandstone is twice as thick.

This formation presents the spectacle of an upheaval immobilized at a particular moment and petrified forever. Hugh Miller wrote:

"The first scene in [Shakespeare's] *The Tempest* opens amid the confusion and turmoil of the hurricane—amid thunders and lightnings, the roar of the wind, the shouts of the seamen, the rattling of cordage, and the wild dash of the billows. The history of the period represented by the Old Red Sandstone seems, in what now forms the northern half of Scotland, to have opened

[1] Hugh Miller, *The Old Red Sandstone* (Boston, 1865; first published in England in 1841), p. 48.
[2] Ibid., p. 217.

in a similar manner. . . . The vast space which now includes Orkney and Loch Ness, Dingwall and Gamrie, and many a thousand square miles besides, was the scene of a shallow ocean, perplexed by powerful currents, and agitated by waves. A vast stratum of water-rolled pebbles, varying in depth from a hundred feet to a hundred yards, remains in a thousand different localities, to testify of the disturbing agencies of this time of commotion." Miller found that the hardest masses in the stratum—"porphyries of vitreous fracture that cut glass as readily as flint, and masses of quartz that strike fire quite as profusely from steel,—are yet polished and ground down into bullet-like forms. . . . And yet it is surely difficult to conceive how the bottom of any sea should have been so violently and so equally agitated for so greatly extended a space . . . and for a period so prolonged, that the entire area should have come to be covered with a stratum of rolled pebbles of almost every variety of ancient rock, fifteen stories' height in thickness."[3]

In the red sandstone an abundant aquatic fauna is embedded. The animals are in disturbed positions. At the period of the past when these formations were composed, "some terrible catastrophe involved in sudden destruction the fish of an area at least a hundred miles from boundary to boundary, perhaps much more. The same platform in Orkney as at Cromarty is strewed thick with remains, which exhibit unequivocally the marks of violent death. The figures are contorted, contracted, curved; the tail in many instances is bent around to the head; the spines stick out; the fins are spread to the full, as in fish that die in convulsions. The Pterichthys[4] shows its arms extended at their stiffest angle, as if prepared for an enemy. The attitudes of all the ichthyolites [any fossil fish] on this platform are attitudes of fear, anger and pain. The remains, too, appear to have suffered nothing from the after-attacks of predaceous fishes; none such seem to have survived. The record is one of destruction at once widely spread and total. . . ."[5]

What agency of destruction could have accounted for "innumerable existences of an area perhaps ten thousand square

---

[3] Ibid., pp. 217–18.
[4] An extinct fishlike animal with winglike projections and with the anterior of the body encased in bony plates.
[5] Miller, *The Old Red Sandstone*, p. 222.

miles in extent [being] annihilated at once"? "Conjecture lacks footing in grappling with the enigma, and expatiates in uncertainty over all the known phenomena of death," wrote Miller.[6]

The ravages of no disease, however virulent, could explain some of the phenomena of this arena of death. Rarely does disease fall equally on many different genera at once, and never does it strike with instantaneous suddenness; yet in the ruins of this platform from ten to twelve distinct genera and many species were involved; and so suddenly did the agency perform its work that its victims were fixed in their first attitude of surprise and terror.

The area of the Old Red Sandstone investigated by Miller comprises one half of Scotland, from Loch Ness to the land's northern extremity and beyond to the Orkney Islands in the north. "A thousand different localities" disclose the same scene of destruction.

An identical picture can be found in many other places all around the world, in similar and dissimilar formations. Of Monte Bolca, near Verona in northern Italy, Buckland wrote: "The circumstances under which the fossil fishes are found at Monte Bolca seem to indicate that they perished suddenly. . . . The skeletons of these fish lie parallel to the laminae of the strata of the calcareous slate; they are always entire, and closely packed on one another. . . . All these fishes must have died suddenly . . . and have been speedily buried in the calcareous sediment then in the course of deposition. From the fact that certain individuals have even preserved traces of colour upon their skin, we are certain that they were entombed before decomposition of their soft parts had taken place."[7]

The same author wrote about the fish deposits in the area of the Harz Mountains in Germany: "Another celebrated deposit of fossil fishes is that of the cupriferous slate surrounding the Harz. Many of the fishes of this slate at Mansfield, Eisleben, etc., have a distorted attitude, which has often been assigned to writhing in the agonies of death. . . . As these fossil fishes maintain the attitude of the rigid stage immediately succeeding death, it follows that they were buried before putrefaction had

[6] Ibid., p. 223.
[7] W. Buckland, *Geology and Mineralogy* (Philadelphia, 1837), p. 101.

commenced, and apparently in the same bituminous mud, the influx of which had caused their destruction."[8]

The story of agony and sudden death and immediate encasing is told by the red sandstone of Scotland; the limestone of Monte Bolca in Lombardy; the bituminous slate of Mansfield in Thuringia; and also by the coal formation of Saarbrücken on the Saar, "the most celebrated deposits of fossil fishes in Europe"; the calcareous slate of Solenhofen; the blue slate of Glaris; the marlstone of Oensingen in Switzerland and of Aix in Provence, to mention only a few of the better-known sites in Europe.

In North America similar strata, "packed full of splendidly preserved fishes" are found in the black limestone of Ohio and Michigan, in the Green River bed of Arizona, the diatom beds of Lompoc, California, and in many other formations.[9]

In cataclysms of early ages fishes died in agony; and the sand and the gravel of the upthrust sea bottom covered the aquatic graveyards.

[8] Ibid., p. 103.
[9] George McCready Price, *Evolutionary Geology and New Catastrophism* (1926), p. 236; J. M. Macfarlane, *Fishes the Source of Petroleum* (1923).

Chapter III

UNIFORMITY

The Doctrine of Uniformity

For over twenty-five years, from the beginning of the French Revolution in 1789 to the Battle of Waterloo in 1815, Europe was in turmoil. France beheaded her king and queen; many revolutionaries in their turn went to the scaffold too. Spain, Italy, Germany, Austria, and Russia became battlefields. The British Isles were in danger of being invaded, and Britain's fleet fought at Trafalgar the tyrant who had sprung up from the revolutionary army. After 1815 there was a universal desire for peace and tranquility. The Holy Alliance was organized; Europe sank into reaction, England into a spirit of conservatism. The abortive revolutionary wave of 1830 did not reach the British Isles.

No wonder that in the climate of reaction to the eruptions of revolution and the Napoleonic Wars the theory of uniformity became popular and soon dominant in the natural sciences. According to this theory, the development of the surface of the globe has been going on through all the ages without any disturbances; the process of very slow change that we observe at present has been the only process of importance from the beginning.

This theory, first advanced by Hutton (1795) and Lamarck (1800), was elevated to its present position as a scientific law by Charles Lyell, a young attorney whose interest in geology was to make him the most influential person in that field, and by Lyell's disciple and friend, Charles Darwin. Darwin built his theory of evolution on Lyell's principle of uniformity. A modern exponent of the theory of evolution, H. F. Osborn, wrote:

22

"Present continuity implies the improbability of past catastrophism and violence of change, either in the lifeless or in the living world; moreover, we seek to interpret the changes and laws of past time through those which we observe at the present time. This was Darwin's secret, learned from Lyell."[1] Lyell built his case with convincing dialectics.

Wind and solar heat and rain little by little crumble the rock in the highlands. Rivers carry the detritus to the sea. The land is lowered by this process, which continues for ages, until it turns a vast region into detritus. Then the massive earth, as if in a slow breathing process, every phase of which requires eons, again slowly rises, the bottom of the sea subsides, and the crumbling of the rock begins all over again. The land comes up in an elevated plateau; the subsequent action of water and wind cuts furrows, and little by little the highland changes into a range of mountain peaks; more eons, and these heights crumble too, wind and rain carrying them grain by grain into the sea; the shallow sea encroaches on the land, then slowly retreats. No great catastrophes intervene to change the face of the earth. Although sporadic volcanic action occurs, Lyell did not consider it to have an effect in changing the face of the earth comparable in importance to that of rivers, wind, and waves of the sea.

What causes the eon-long process of elevation and subsidence has not been determined. Naturalists of the eighteenth century claimed to have observed a minute gradual change in the level of the Gulf of Bothnia in the Baltic Sea in relation to the coast line. Similar processes in past geological ages must have brought about all the changes on the earth: the majestic mountains that rose and others that were levelled, the seacoast that moved in a slow rhythm back and forth, and the earth mantle that was redistributed by rain and wind. According to the theory of uniformity, no process took place in the past that is not taking place at present; and not only the nature but also the intensity of physical phenomena of our age are the criteria of what could have happened in the past.

Since the theory of uniformity is still taught in all places of learning, and to question it is heresy, it is pertinent to reproduce here some of Lyell's original statements, made in his *Principles*

[1] H. F. Osborn, *The Origin and Evolution of Life* (1917), p. 24.

of Geology; they served as a manifesto or credo for all his followers, whether called uniformists or evolutionists. Lyell wrote:

"It has been truly observed that when we arrange the known fossiliferous formations in chronological order, they constitute a broken and defective series . . . we pass, without any intermediate gradations from systems of strata which are horizontal, to other systems which are highly inclined—from rocks of peculiar mineral composition to others which have a character wholly distinct—from one assemblage of organic remains to another, in which frequently nearly all the species, and a large part of the genera, are different. These violations of continuity are so common as to constitute in most regions the rule rather than the exception, and they have been considered by many geologists as conclusive in favour of sudden revolutions in the inanimate and animate world."[2]

Thus he acknowledged that the surface of the globe has the appearance of having been subjected to great and violent sudden changes, but he believed that the record is incomplete and that the major part of the evidence is lost. "In the solid framework of the globe we have a chronological chain of natural records, many links of which are wanting."[3] To make this plausible, Lyell cited an example from human affairs. If a census were taken every year in sixty provinces, changes in the population would appear to be very gradual; but if the census were taken every year in a different province, and in only one, the change in the population of each province between the visits of the census takers at sixty-year intervals would be very great. Lyell maintained that this was the way geological deposits were made.

The theory of uniformity, or of gradual changes in the past measured by the extent of changes observed in the present, has, as Lyell admitted, no positive evidence in the incomplete record of the earth's crust; consequently the theory, building on *argumentum ex silentio*, or argument by default, required further analogies.

"Suppose we have discovered two buried cities at the foot of Vesuvius, immediately superimposed upon each other with

[2] Sir Charles Lyell, *Principles of Geology* (12th ed.; 1875), I, 298.
[3] Ibid., p. 299.

24

a great mass of tuff and lava intervening. . . . An antiquary [archaeologist] might possibly be entitled to infer, from the inscriptions on public edifices, that the inhabitants of the inferior and older city were Greeks, and those of the modern town Italians. But he would reason very hastily if he also concluded from these data, that there had been a sudden change from the Greek to the Italian language in Campania. But if he afterwards found three buried cities, one above the other, the intermediate one, being Roman . . . he would then perceive the fallacy of his former opinion, and would begin to suspect that the catastrophes, by which the cities were inhumed, might have no relation whatever to the fluctuations in the language of the inhabitants; and that, as the Roman tongue had evidently intervened between the Greek and Italian, so many other dialects may have been spoken in succession, and the passage from the Greek to the Italian may have been very gradual. . . ."[4]

This often-reprinted passage is an unfortunate example, for, in order to prove that there had been no violent changes, Lyell chose to present a picture of violent catastrophes: the strata are separated by layers of lava. This is also the picture presented in so many geological surveys. To use this example as a proof of uniformity is a flight of dialectics.

The comparison is followed by an accusation that is all the more vigorous because of the inadequacy of the example which is called on to substitute for geological evidence. Lyell said:

"It appeared clear that the earlier geologists had not only a scanty acquaintance with existing changes [caused by wind, flowing water, etc.], but were singularly unconscious of the amount of their ignorance. With the presumption naturally inspired by this unconsciousness, they had no hesitation in deciding at once that time could never enable the existing powers of nature to work out changes of great magnitude, still less such important revolutions as those which are brought to light by geology."[5]

And he proceeded:

"Never was there a dogma more calculated to foster indolence, and to blunt the edge of curiosity, than this assumption of the discordance between the ancient and existing causes of change. It produced a state of mind unfavourable in the highest

[4] Ibid., p. 316. [5] Ibid., p. 317.

degree to the candid reception of the evidence of those minute but incessant alterations which every part of the earth's surface is undergoing."[6]

At first the tone of this pleading for the then unorthodox theory of uniformity was defensive; the position was unsupported by sufficient evidence. Then, as though a few analogies to human situations were so strong that they could substitute for the defective record of nature, the tone changed and became uncompromising.

"For this reason all theories are rejected which involve the assumption of sudden and violent catastrophes and revolutions of the whole earth, and its inhabitants—theories which are restrained by no reference to existing analogies, and in which a desire is manifested to cut, rather than patiently to untie, the Gordian knot."[7]

Notwithstanding the strong language employed, the scientific principle which insists that whatever does not occur at the present time has not occurred in the past is a self-imposed limitation. Rather than a principle in science, it is a statute of faith. And Lyell ended his famous chapter accordingly, with an appeal for faith and with a precept for believers:

"If he [the student] finally believes in the resemblance or identity of the ancient and present systems of terrestrial changes, he will regard every fact collected respecting the causes in diurnal action as affording him a key to the interpretations of some mystery in the past."[8]

The Hippopotamus

The hippopotamus inhabits the larger rivers and marshes of Africa; it is not found in Europe or America save in zoological gardens where specimens of it wallow most of the time in pools, submerging their huge bodies in muddy water. Next to the elephant it is the largest of the land animals. Bones of hippopotami are found in the soil of Europe as far north as Yorkshire in England.

Lyell gave the following explanation for the presence of the hippopotamus in Europe:

[6] Ibid., p. 318.   [7] Ibid.   [8] Ibid., p. 319.

"The geologist . . . may freely speculate on the time when herds of hippopotami issued from North African rivers, such as the Nile, and swam northward in summer along the coasts of the Mediterranean, or even occasionally visited islands near the shore. Here and there they may have landed to graze or browse, tarrying awhile, and afterwards continuing their course northward. Others may have swum in a few summer days from rivers in the south of Spain or France to the Somme, Thames, or Severn, making timely retreat to the south before the snow and ice set in.'[1]

An Argonaut expedition of hippopotami from the rivers of Africa to the isles of Albion sounds like an idyll.

In the Victorian cave near Settle, in west Yorkshire 1450 feet above sea level, under twelve feet of clay deposit containing some well-scratched boulders, were found numerous remains of the mammoth, rhinoceros, hippopotamus, bison, hyena, and other animals.

In northern Wales in the Vale of Clwyd, in numerous caves remains of the hippopotamus lay together with those of the mammoth, the rhinoceros, and the cave lion. In the cave of Cae Gwyn in the Vale of Clwyd, "during the excavations it became clear that the bones had been greatly disturbed by water action." The floor of the cavern was "covered afterwards by clays and sand containing foreign pebbles. This seemed to prove that the caverns, now 400 feet [above sea level] must have been submerged subsequently to their occupation by the animals and by man. . . . The contents of the cavern must have been dispersed by *marine* action during the great submergence in mid-glacial times, and afterwards covered by marine sands . . ." writes H. B. Woodward.[2]

Hippopotami not only travelled during the summer nights to England and Wales, but also climbed hills to die peacefully among other animals in the caves, and the ice, approaching softly, tenderly spread little pebbles over the travellers resting in peace, and the land with its hills and caverns in a slow lullaby movement sank below the level of the sea and gentle streams caressed the dead bodies and covered them with rosy sand.

[1] Charles Lyell, *Antiquity of Man* (1863), p. 180.
[2] H. B. Woodward, *Geology of England and Wales* (2nd ed.; 1887), p. 543.

Three assumptions were made by the exponents of uniformity: Sometime not long ago the climate of the British Isles was so warm that hippopotami used to visit there in summer; the British Isles subsided so much that caves in the hills became submerged; the land rose again to its present height—and all this without any action of a violent nature.

Or was it, perchance, a mountain-high wave that crossed the land and poured into the caves and filled them with marine sand and gravel? Or did the ground submerge and then emerge again in some paroxysm of nature in which the climate also changed? Did the animals run away at the sign of the approaching catastrophe, and did the trespassing sea follow and suffocate them in the caves that were their last refuge and became the place of their burial? Or did the sea sweep them from Africa, throw them in heaps on the British Isles and in other places, and cover them with earth and marine debris? The entrances to some caves were too narrow and the caves themselves too "shrunk" (contracted) to have been places of refuge for such huge animals as hippopotami and rhinoceroses. Whichever of these answers or surmises is correct, and whether the hippopotami lived in England or were thrown there by the ocean, whether they sought refuge in caves or the caves are but their graves, their bones on the British Isles, as also on the bottom of the seas surrounding these islands, are signs of some great natural change.

Icebergs

The theory that rejected the occurrence of catastrophic events in the past was incompatible with the then prevailing teaching, which ascribed the distribution of drift (the deposit of rock debris, clay, and organic material that covers continental areas) and of erratic boulders to the action of water in the form of great tidal waves breaking upon the continents. A slow-moving source, able to do the same work, but in a longer time, had to be found. Lyell assumed that icebergs transferred rocks over the expanse of the sea. Icebergs are broken-off parts of glaciers that descend from the mountainous coasts to the sea. Mariners in northern waters have observed icebergs with pieces of rock attached to them. And if we think of the enormity of past geological epochs and multiply the action of icebergs as carriers of earth and rocks by the time elapsed, we may explain, so argued Lyell, the presence of erratic boulders as well as of till and gravel on land.

Erratic boulders are found far from the seashore: Lyell taught that the land was submerged and icebergs travelling over it dropped their load of stones; later the land emerged with the stones on it. Erratic boulders are found on the mountains; therefore, these mountains were under shallow water when icebergs carrying stones from other regions dropped them on the summits. In order to explain in this manner the provenience of erratic boulders, it was necessary to submerge large parts of continents in rather recent times.

In some places erratic boulders are distributed in a long string —as in the Berkshires. Icebergs could not have acted as intelligent carriers, and Lyell must have felt the weakness of his theory on this point. The only alternative known at that time was that of a tidal wave. But Lyell abhorred catastrophes. He detested them alike in the political life of Europe and in nature. Characteristically, his autobiography begins with this description of the most vivid memory of his early childhood:

"I was four and a half when an event happened which was not likely to be forgotten." His family travelled in two carriages a stage and a half from Edinburgh. "On a narrow road, with a steep brae above, and an equally precipitous one below, and no parapet on the roadside, a flock of sheep jumped down into the road, and frightened the horses [of the other carriage]. Away they ran, and with the chaise, man, horses and all, disappeared clean out of sight, over the brae in an instant." There was a rescue through the broken pane of glass, a little blood ran, and somebody fainted.[1] It left the first and strongest impression of his childhood in the memory of the author of the theory of uniformity.

Darwin in South America

Charles Darwin, who had previously dropped his medical studies at Edinburgh University, upon his graduation in theology from Christ College, Cambridge, went in December 1831

[1] Charles Lyell, *Life, Letters and Journals* (1881), I, 2.

as a naturalist on the ship *Beagle*, which sailed around the world on a five-year surveying expedition. Darwin had with him the newly published volume of Lyell's *Principles of Geology* that became his Bible. On this voyage he wrote his *Journal*, the second edition of which he dedicatgd to Lyell.

This round-the-world voyage was Darwin's only field-work experience in geology and paleontology, and he drew on it all his life long. He wrote later that these observations served as the "origin of all my views." His observations were made in the Southern Hemisphere and more particularly in South America, a continent that had attracted the attention of naturalists since the exploration travels of Alexander von Humboldt (1799–1804). Darwin was impressed by the numerous assemblages of fossils of extinct animals, mostly of much greater size than species now living; these fossils spoke of a flourishing fauna that suddenly came to its end in a recent geological age. He wrote under January 9, 1834, in the *Journal* of his voyage:

"It is impossible to reflect on the changed state of the American continent without the deepest astonishment. Formerly it must have swarmed with great monsters: now we find mere pigmies, compared with the antecedent, allied races."

He proceeded thus: "The greater number, if not all, of these extinct quadrupeds lived at a late period, and were the contemporaries of most of the existing sea-shells. Since they lived, no very great change in the form of the land can have taken place. What then, has exterminated so many species and whole genera? The mind at first is irresistibly hurried into the belief of some great catastrophe; but thus to destroy animals, both large and small, in Southern Patagonia, in Brazil, on the Cordillera of Peru, in North America up to Behring's [Bering's] Straits, *we must shake the entire framework of the globe*."

No lesser physical event could have brought about this wholesale destruction, not only in the Americas but in the entire world. And such an event being beyond consideration, Darwin did not know the answer. "It could hardly have been a change of temperature, which at about the same time destroyed the inhabitants of tropical, temperate, and arctic latitudes on both sides of the globe."

Certainly it could not have been man in the role of the destroyer; and were he to attack all large animals, would he also be the cause of extinction "of the many fossil mice and other small quadrupeds?" Darwin asked.

"No one will imagine that a drought ... could destroy every individual of every species from Southern Patagonia to Behring's Straits. What shall we say of the extinction of the horse? Did those plains fail of pasture, which have since been overrun by thousands and hundreds of thousands of the descendants of the stock introduced by the Spaniards?" Darwin concluded: "Certainly, no fact in the long history of the world is so startling as the wide and repeated exterminations of its inhabitants."[1] Out of Darwin's embarrassment grew the ideas of extinction of species as a prelude to natural selection.

[1] Charles Darwin, *Journal of Researches into the Natural History and Geology of the Countries Visited During the Voyage of H.M.S. Beagle Round the World*, under date of January 9, 1834.

Chapter IV

ICE

The Birth of the Ice Age Theory

In 1836, Louis Agassiz, a young Swiss naturalist, went with Professor Jean Charpentier, another naturalist, to an alpine glacier to demonstrate to him the fallacy of the new idea that an ice sheet once covered a large part of Europe. Four years before, a teacher in a small-town forestry school, A. Bernardi, had written: "Once the polar ice reached as far as the southern limit of the district which is still marked by the erratics."[1] A botanist, C. Schimper, had come upon the same idea, probably independently, and coined the term *die Eiszeit*; he had succeeded in winning Charpentier to the hypothesis. At the edge of the glacier, Agassiz, who came as sceptic, was himself converted; he became the chief apostle of the new theory. He built a hut on the glacier of Aar and lived in it, so that he could observe the movements of the ice, and thereby attracted the attention of naturalists and curiosity seekers all over Europe.

The study of the glaciers in the Alps revealed that glacial ice may move by its own weight a few feet daily; it actually transports stones by carrying and pushing them. Some loose rocks are shoved aside to form lateral moraines; some are pushed by the advancing front of the ice to form terminal moraines. When the glacier melts and retreats, the loose rocks remain where they were at the time of the greatest expansion of the glacier. Agassiz assumed that the erratic boulders on the Jura Mountains had been carried there by ice from the Alps and that the trains of

[1] A. Bernardi, "Wie kamen die aus dem Norden stammenden Fels-bruchstücke und Geschiebe, welche man in Norddeustchland und den benachbarten Ländern findet, an ihre gegenwärtigen Fundorte?" *Jahrbuch für Mineralogie, Geognosie und Petrefactenkunde*, III (1832), 57–67.

boulders in northern Europe and America had been formed by the gigantic glaciers that, sometime in the past, covered large parts of these continents. He also concluded that the drift had been brought and left by the ice sheet. Ice scratched the underlying rock with the help of flint and other fragments of hard stone it retained in its grasp; and it polished the rocky floors of slopes and valleys, and excavated the beds of lakes.

Agassiz made his conclusions with respect to other parts of the world on the basis of observations limited to Switzerland and its surroundings. He thought that if he could convert two of the leading geologists, Buckland, the author of *Reliquiae diluvianae*, and Murchison, to the ice-age theory and thus win their support, the task of gaining recognition for it would become much easier. Agassiz went to the British Isles. In later years, as his widow described it, "recalling the scientific isolation in which he then stood, opposed as he was to all the prominent geologists of the day, he said: 'Among the older naturalists, only one stood by me: Dr. Buckland, Dean of Westminster. . . . We went first to the Highlands of Scotland, and it is one of the delightful recollections of my life that as we approached the castle of the Duke of Argyll, standing in a valley not unlike some of the Swiss valleys, I said to Buckland: "Here we shall find our first traces of glaciers"; and, as the stage entered the valley, we actually drove over an ancient terminal moraine, which spanned the opening of the valley'."[2] It was a setting for a revelation. Agassiz won a follower.

A few weeks later, on November 4, 1840, Agassiz read a paper before the Geological Society of London, summarizing the excursion in the light of the Ice Age theory, and Buckland, who was then president of the society, followed with a paper of his own on the same subject. Even before the meeting he had written to Agassiz of the success of his missionary work; "Lyell has adopted your theory in toto!!! On my showing him a beautiful cluster of moraines, within two miles of his father's house, he instantly accepted it, as solving a host of difficulties that have all his life embarrassed him."[3] Lyell, too, agreed to read a paper less than three weeks after this episode, on the day

[2] *Louis Agassiz, His Life and Correspondence*, ed. Elizabeth Cary Agassiz (1893), I, 307.

[3] Ibid., I, 309.

following the Agassiz and Buckland lectures. In this paper, hastily prepared, he explained the moraines in Great Britain in the light of Agassiz's teachings.

At the November 4 meeting of the society, Murchison "attempted an opposition" but, in words of Agassiz, "did not produce much effect." He added: "Dr. Buckland was truly eloquent."

That same year (1840) Agassiz published his theory in a work entitled *Etudes sur les glaciers*. He wrote:

"The surface of Europe previously adorned with tropical vegetation and populated by herds of huge elephants, enormous hippopotami, and gigantic carnivora, was suddenly buried under a vast mantle of ice, covering plains, lakes, seas, and plateaux. Upon the life and movement of a vigorous creation fell the silence of death. Springs vanished, rivers ceased flowing, the rays of the sun, rising upon this frozen shore (if, indeed, they reached it), encountered only the breath of winter from the north and the thunder of crevasses as they opened up across the surface of this icy sea."[4]

Agassiz regarded the inception and the termination of the Ice Age as catastrophic events. He believed that mammoths in Siberia were suddenly caught in the ice that spread swiftly over the larger part of the globe. He expressed the belief that repeated global catastrophes were accompanied by a fall in the temperature of the globe and its atmosphere, and that glacial ages, of which the earth experienced more than one, were terminated each time by renewed igneous activity in the interior of the earth (*éruptions de l'interieur*). Thus he maintained that the western Alps had risen very recently, at the *end* of the last Ice Age, and were younger than the carcasses of mammoths in Siberia, the flesh of which is still edible: these animals, he thought, had been killed at the *beginning* of the Ice Age.[5] With the renewal of igneous activity, the ice cover melted, great floods ensued, the mountains and lakes in Switzerland and in many other places were formed, and the relief map of the world was generally changed.

It is often said that Agassiz added from half a million to a million years to the recent history of the world by inserting the

[4] Louis Agassiz, *Etudes sur les glaciers* (1840), p. 314.
[5] Ibid., pp. 304–29.

Great Ice Age between the Tertiary, or the age of mammals, and the Recent (comprising the Late Stone Age and historical times). It should be borne in mind, however, that the million-year span for the Ice Age is Lyell's estimate, and he interpreted Agassiz's theory in the spirit of uniformity.[6]

The theory of a continental ice cover was acceptable to Lyell. He agreed to it, satisfied to go no farther for his proof than two miles from his home. He realized that floating icebergs could not explain the phenomenon of drift and erratic boulders in all places. The only alternative had been the waves of translation, or tidal waves travelling on land, but this was completely catastrophic. Now, with the continental ice theory, he felt he had the correct solution if the catastrophic aspect of the theory, as originally suggested by Agassiz, a follower of Cuvier, was eliminated. It was not yet asked what produced such a cover.

On the Russian Plains

Soon after the historic meeting at which the Ice Age theory was accepted by the majority of the members of the Geological Society, R. I. Murchison went to Russia, where he had been invited by Czar Nicholas I to make a geological survey of the empire. Out of this survey grew recognition of the Permian System; the Permian, Silurian, and the Devonian, also first recognized by Murchison (Devonian in collaboration with Sedgwick), constitute three of the great divisions in the modern concept of early geological ages. For many months Murchison crossed the latitudes and longitudes of Russia, carefully observing the erratic boulders strewn over the great Russian plains and rechecking the validity of Agassiz's theory. In Finland and the northern Russian provinces he found very large blocks; but they diminished in size the farther south one went, which pointed to the action of water, a tide that came down from the north or northwest, spreading rock fragments along its way. He also observed that erratic boulders in the Carpathian Mountains were not of local but of Scandinavian origin.

[6] Lyell borrowed the estimate of a million-year span of time for the Ice Age from J. Croll, who needed this length of time for his astronomical theory of glacial periods, a theory long since abandoned.

Of the drift, or "the piles of stone, sand, clay and gravel which are spread out in such enormous masses over the low countries of Russia, Poland, and Germany," Murchison expressed the conviction that "a vast portion, by far the greater part . . . has been transported by aqueous action, consequent of powerful waves of translation and currents occasioned by relative and often paroxysmal changes of the level of sea and land."[1] Whatever may have been the cause of the irruption of the sea, such aqueous debacles "with the help of ice floes" produced the drift.

"Seeing that there are no mountains whatever from which a glacier can ever have been propelled in southern Sweden, Finland, or north-eastern Russia, and yet that these regions are powerfully abraded, scored and polished," Murchison came to the conclusion that effects so extensively developed over such flat countries must have resulted from an irrupting sea that also left behind enormous masses of debris and rolled stone.

Murchison "rejected the application of the terrestrial glacier theory to Sweden, Finland, north-eastern Russia, and the whole of northern Germany—in short to all the low countries of Europe."[2] He agreed that in mountainous northern Scandinavia and Lapland arctic glaciers formerly did exist. Ice floes descending from these glaciers carried angular broken stones over land covered by sea and dropped them on top of the drift created by the irruption of the sea.

Murchison called attention to the fact that "Siberia is entirely free from erratic blocks, though environed on three sides by high mountains."[3]

He required the aid of icebergs detached from the glaciers to "account for certain superficial phenomena," but he confidently maintained that "aqueous detrital conditions will best account for the great diffusion of drift over the surface of the globe, and at the same time explain the very general striation and abrasion of the rocks, at low as well as high levels, in numerous parallels of latitude."[4]

In his later years, Murchison, without retracting any of his observations and conclusions made in Russia, admitted in a

[1] R. I. Murchison, *The Geology of Russia in Europe and the Ural Mountains*, I (London, 1845), 553.
[2] Ibid., p. 554. [3] Ibid. [4] Ibid.

letter to Agassiz that he regretted his early opposition to the Ice Age theory. On the other hand, marine deposits of recent age were found in large areas of European and Asiatic Russia. In the Caspian Sea, which stretches between southern Russia and Persia, live seals related to the seals of the Arctic Ocean. It is concluded that the polar sea spread and established a connection with the Caspian Sea, and this in Recent time.

"Since the ice withdrew, the Arctic Ocean has spread over large areas of northern Russia and in many places has left marine deposits on the glacial drift as well as on the firmer rocks. The Arctic water spread also over the Obi Basin far to the south and established connections with the Caspian Sea, at which time the progenitors of the present seals of the Caspian rocky islands migrated thither to become stranded when the waters withdrew."[5]

Ice Age in the Tropics

In 1865, Agassiz went to equatorial Brazil, one of the hottest places in the world, where he found all the signs he ascribed to the action of ice. Now even those who had previously agreed with him became distressed. An ice cover in the tropics, on the very equator? There were drift accumulations, and scratched rocks, and erratic boulders, and fluted valleys, and the smooth surface of tillite (rock formed of consolidated till), so there must have been ice to carry and polish, and the region must have gone through an ice period. What could have caused a tropical region to be covered by ice several thousand feet thick?

Abundant vestiges of an ice age were likewise found in British Guiana, another of the hottest places on earth.

Soon the same word came from equatorial Africa; and what appeared even more strange, the marks there indicated not only that equatorial Africa and Madagascar had been under a sheet of ice but that the ice had moved, spreading *from* the equator toward the higher latitudes of the Southern Hemisphere, or in the wrong direction.

Then vestiges of an ice age were discovered in India, and there, too, the ice had moved *from* the equator, and not merely

[5] G. D. Hubbard, *The Geography of Europe* (1937), p. 47.

toward higher latitudes, but uphill, from the lowland up the foothills of the Himalayas.

On reconsideration, the vestiges of ice in equatorial regions were ascribed to a different ice age that had taken place not thousands but many millions of years ago. Today the glacial phenomena in the tropics and in the Southern Hemisphere are ascribed, in the main, to the Permian Age, a much earlier period than the recent Ice Age. "The most remarkable feature of the Permian glaciation is its distribution," writes C. O. Dunbar of Yale University. "South America bears evidence of glaciation in Argentina and southeastern Brazil, even within 10° of the equator. In the northern hemisphere, peninsular India, within 20° of the equator, was the chief scene of glaciation, with the ice flowing north [or from the tropics to higher latitudes]."[1] "The icecap covered practically all of southern Africa up to at least latitude 22°S and also spread to Madagascar."[2]

Even if the phenomenon took place very long ago, an ice cover thousands of feet thick in the hottest places of the world is a challenging enigma. R. T. Chamberlin says: "Some of these huge ice sheets advanced even into the tropics, where their deposits of glacier-borne debris, hundreds of feet in thickness, amaze the geologists who see them. No satisfactory explanation has yet been offered for the extent and location of these extraordinary glaciers. . . . Glaciers, almost unbelievable because of their location and size, certainly did not form in deserts. . . ."[3]

Greenland

Greenland is the contemporary example of what, according to the Ice Age theory, happened to a large part of the world in times past. Greenland belongs to the great archipelago that crowns northeastern Canada, though it is sometimes regarded as a part of Europe. It is the largest island in the world, if we consider Antarctica and Australia as continents. The island is 1660 miles long, largely within the Arctic Circle, reaching the

[1] C. O. Dunbar, *Historical Geology* (1949), pp. 298–99.   [2] Ibid., p. 298.
[3] R. T. Chamberlin, "The Origin and History of the Earth" in *The World and Man*, ed. F. R. Moulton (1937), p. 80.

northern latitude of 83° 39'. Of its 840,000 square miles of surface, over 700,000 are covered with an immense mountain of ice that leaves free only the coastal fringes. The thickness of the ice is measured by listening to the echo that comes from the bedrock when a detonation is set off on top of the ice. It is found to be over six thousand feet thick.

"For a long time it was the belief of many that a large region in the interior of Greenland was free from ice, and was perhaps inhabited. It was in part to solve this problem that Baron [N. A. E.] Nordenskjöld set out upon his expedition of 1883."[1] He ascended from the icecap from Disko Bay (latitude 69°) and went eastward for eighteen days across the ice field. "Rivers were flowing in channels upon the surface like those cut on land . . . only that the pure blue of the ice-walls was, by comparison, infinitely more beautiful. These rivers were not, however, perfectly continuous. After flowing for a distance in channels on the surface, they, one and all, plunged with deafening roar into some yawning crevasse, to find their way to the sea through subglacial channels. Numerous lakes with shores of ice were also encountered."

"On bending down the ear to the ice," wrote the explorer, "we could hear on every side a peculiar subterranean hum, proceeding from rivers flowing within the ice; and occasionally a loud single report like that of a cannon gave notice of the formation of a new glacier-cleft. . . . In the afternoon we saw at some distance from us a well-defined pillar of mist which, when we approached it, appeared to rise from a bottomless abyss, into which a mighty glacier-river fell. The vast roaring watermass had bored for itself a vertical hole, probably down to the rock, certainly more than 2000 feet beneath, on which the glacier rested."[2]

The Ice Age survived in Greenland. This arctic island reveals how vast continental areas looked in the past. However, it does not explain how ice could have covered British Guiana or Madagascar in the tropics. And what is no less surprising, the northern part of Greenland, according to the concerted opinion of glaciologists, was never glaciated. "Probably, then as now, an exception was the northernmost part of Greenland; for it seems a rule that the most northernly lands are not, and never

[1] Wright, *The Ice Age in North America*, p. 75.   [2] Ibid.

39

were, glaciated," writes the polar explorer Vilhjalmur Stefansson.[3] "The islands of the Arctic Archipelago," writes another scientist, "were never glaciated. Neither was the interior of Alaska."[4] "It is a remarkable fact that no ice mass covered the low lands of northern Siberia any more than those of Alaska," wrote James D. Dana, the leading American geologist of the last century.[5] In northern Siberia and on polar islands in the Arctic Ocean spires of rock were observed that would certainly have been broken off if an ice cover had moved over those parts.[6]

Bones of Greenland reindeer have been found in southern New Jersey and southern France, and bones of Lapland reindeer in the Crimea. This was explained as due to the invasion of ice and the retreat of northern animals to the south. The hippopotamus was found in France and England and the lion in Alaska. To explain similar occurrences, an inter-glacial period was introduced into the scheme: the land was warmed up and the southern animals visited northern latitudes. And since the change from one fauna to another took place repeatedly, four glacial periods with three interglacial were generally counted, though the number of periods is not consistent with all lands or with all investigators.

But why the polar lands were not glaciated during the Ice Age was never explained. Greenland presents still another enigma in the preceding formations, those of the Tertiary Age. In the 1860s, O. Heer of Zurich published his classical work on the fossil plants of the Arctic; he identified the plant remains of the northern parts of Greenland as magnolia and fig trees, among other species.[7] Forests of exotic trees and groves of juicy subtropical plants grew in the land that lies deep in the cold Arctic and is immersed yearly in a continuous polar night of six month's duration.

Corals of the Polar Regions

Spitsbergen in the Arctic Ocean is as far north from Oslo in Norway as Oslo is from Naples. Heer identified 136 species of

[3] V. Stefansson, *Greenland* (1942), p. 4.
[4] R. F. Griggs, *Science*, XCV (1942), 2473.
[5] Dana, *Manual of Geology* (4th ed.), p. 977.
[6] Whitley, *Journal of the Philosophical Society of Great Britain*, XII, 55.
[7] O. Heer, *Flora Artica Fossilis: Die fossile Flora der Polarländer* (1868).

fossil plants from Spitsbergen (78° 56′ north latitude), which he ascribed to the Tertiary Age. Among the plants were pines, firs, spruces, and cypresses, also elms, hazels, and water lilies.

At the nothernmost tip of Spitsbergen Archipelago, a bed of black and lustrous coal twenty-five to thirty feet thick was found; it is covered with black shale and sandstone encrusted with fossilized land plants. "When we remember that this vegetation grew luxuriantly within 8° 15′ of the North Pole, in a region which is in darkness for half of the year and is now almost continuously buried under snow and ice, we can realize the difficulty of the problem in the distribution of climate which these facts present to the geologist."[1]

There must have been great forests on Spitsbergen to produce a bed of coal thirty feet thick. And even if Sptisbergen, almost one thousand miles inside the Arctic Circle, for some unknown reason had the warm climate of the French Riviera on the Mediterranean, still these thick forests could not have grown there, because the place is six months in continuous night. The rest of the year the sun stands low over the horizon.

Not only fossil trees and coal but corals, too, were found there. Corals grow only in tropical water. In the Mediterranean, in the climate of Egypt or Morocco, it is too cold for them. But they grew in Spitsbergen. Today large formations of coral covered with snow can be seen. It does not solve the problem of their deposition, if they were formed in an older geological epoch.

At some time in the remote past corals grew and are still found on the entire fringe of polar North America—in Alaska, Canada, and Greenland.[2] In later times (Tertiary) fig palms bloomed within the Arctic Circle; forests of *Sequoia gigantea*, the giant tree of California, grew from Bering Strait to north of Labrador. "It is difficult to imagine any possible conditions of climate in which these plants could grow so near the pole, deprived of sunlight for many months of the year."[3]

It is usually said that in ages past the climate all over the world was the same, or that a characteristic of the "warm

[1] Archibald Geikie, *Text-Book of Geology* (1882), p. 869.
[2] Dunbar, *Historical Geology*, pp. 162, 194.
[3] D. H. Campbell, "Continental Drift and Plant Distribution," *Science*, January 16, 1942.

periods which have formed the major part of geological time was the small temperature difference between equatorial and polar regions." To this C. E. P. Brooks, in his book, *Climate through the Ages*, says: "So long as the axis of rotation remains in nearly its present position relative to the plane of the earth's orbit round the sun, the outer limit of the atmosphere in tropical regions must receive more of the sun's heat than [in] the middle latitudes, and [in] the middle latitudes more than [in] the polar regions; this is an invariable law. . . . It is much more difficult to think of a cause which will raise the temperature of polar regions by some 30°F. or more, while leaving that of equatorial regions almost unchanged."[4]

The continent of Antarctica is larger than Europe, European Russia included. It has not a single tree, not a single bush, not a single blade of grass. Very few fungi have been found. Reports of polar explorers indicate that no land animals larger than insects have been seen, and these insects are exceedingly few and degenerate. Penguins and sea gulls come from the sea. Storms of great velocity circle the Antarctic most of the year. The greatest part of the continent is covered with ice that in some places descends into the ocean.

E. H. Shackleton, during his expedition to Antarctica in 1907–9, found fossil wood in the sandstone of a moraine at latitude 85° 5'. He also found erratic boulders of granite on the slopes of Mount Erebus, a volcano. Then he discovered seven seams of coal, also at about latitude 85°. The seams are each between three and seven feet thick. Associated with the coal is sandstone containing coniferous wood.[5]

Antarctica, too, must have had great forests in the past.

It often appears that the historian of climate has chosen a field as hard to master as it is to square the circle. It seems sometimes that the history of climate is a collection of unsolved, even unsolvable, questions. Without drastic changes in the position of the terrestrial axis or in the form of the orbit or both, conditions could not have existed in which tropical plants flourished

[4] C. E. P. Brooks, *Climate through the Ages* (1949), p. 31.
[5] Shackleton, *The Heart of the Antarctic*, II, 314, 316, 319, 323, and photographs opposite pp. 293, 316. According to Chamberlin, coal is found only two hundred miles from the South Pole.

in polar regions. If anyone is not convinced of this, he should try to cultivate coral at the North Pole.

Whales in the Mountains

In bogs covering glacial deposits in Michigan, skeletons of two whales were discovered. Whales are marine animals. How did they come to Michigan in the post-glacial epoch? Whales do not travel by land. Glaciers do not carry whales, and the ice sheet would not have brought them to the middle of a continent. Besides, the whale bones were found in *post*-glacial deposits. Was there a sea in Michigan *after* the glacial epoch, only a few thousand years ago?

In order to account for whales in Michigan, it was conjectured that in the post-glacial epoch the Great Lakes were part of an arm of the sea. At present the surface of Lake Michigan is 582 feet above sea level.

Bones of whale have been found 440 feet above sea level, north of Lake Ontario; a skeleton of another whale was discovered in Vermont, more than 500 feet above sea level;[1] and still another in the Montreal-Quebec area, about 600 feet above sea level.[2]

Although the Humphrey whale and beluga occasionally enter the mouth of the St. Lawrence, they do not climb hills. To account for the presence of whales in the hills of Vermont and Montreal, at elevations of 500 and 600 feet, requires the lowering of the land to that extent. Another solution would be for an ocean tide, carrying the whales, to have trespassed upon the land. In either case herculean force would have been required to push mountains below sea level or to cause the sea to irrupt, but the latter explanation is clearly catastrophic. Therefore the accepted theory is that the land in the region of Montreal and Vermont was depressed more than 600 feet by the weight of ice and kept in this position for a while after the ice melted.

But along the coast of Nova Scotia and New England stumps of trees stand in water, telling of once forested country that became submerged. And opposite the mouths of the St. Lawrence and the Hudson rivers are deep canyons stretching for

[1] Dana, *Manual of Geology*, p. 983.   [2] Dunbar, *Historical Geology*, p. 453.

43

hundreds of miles into the ocean. These indicate that the land became sea, being depressed in post-glacial times. Then did both processes go on simultaneously, in neighbouring areas, here up, there down?

A species of Tertiary whale, *Zeuglodon*, left its bones in great numbers in Alabama and other Gulf States. The bones of these creatures covered the fields in such abundance and were "so much of a nuisance on the top of the ground that the farmers piled them up to make fences."[3] There was no ice cover in the Gulf States; then what had caused the submergence and emergence of the land there?

The ocean coast, not only of the area covered by ice, but all the way from Maine to Florida, was at one time submerged and then uplifted. Reginald A. Daly of Harvard wrote: "Not long ago in a geological sense, the flat plain from New Jersey to Florida was under the sea. At that time the ocean surf broke directly on the Old Appalachian Mountains. . . . The wedge-like mass of marine sediments was then uplifted and cut into by rivers, giving the Atlantic Coastal Plain of the United States. Why was it uplifted? To the westward are the Appalachians. The geologist tells us of the stressful times when a belt of rocks extending from Alabama to Newfoundland, was jammed, crumpled, thrust together, to make this mountain system. Why? How was it done? In former times the sea flooded the region of the Great Plains from Mexico to Alaska, and then withdrew. Why this change?"[4]

In Georgia marine deposits occur at altitudes of 160 feet and in northern Florida at altitudes of "at least 240 feet." Walrus is found in Georgian deposits. "Pleistocene [Ice Age] marine features are present along the Gulf coast east of the Mississippi River, in some places at altitudes that may exceed 200 feet."[5] In Texas mammalian land animals of the Ice Age are found in marine deposits. These areas were not covered by the ice which, advancing from the north, reached only as far as Pennsylvania.

A marine deposit overlies the seaboard of northeastern states and the Arctic coast of Canada; in this deposit walrus, seals and

[3] George McCready Price, *Common-sense Geology* (1946), pp. 204–5.
[4] R. A. Daly, *Our Mobile Earth* (1926), p. 90.
[5] R. F. Flint, *Glacial Geology and the Pleistocene Epoch* (1947), pp. 294–95.

at least five genera of whales are found. Marine deposits on land "identified with both glacial and interglacial ages," or containing animals of Arctic and of temperate latitudes, "exist along both Arctic and Pacific coasts in places extending more than 200 miles inland."[6]

The change in land elevation in the region previously covered by ice is ascribed to the removal of the ice cover that weighed down the earth's crust; but what changed the elevation of other areas outside the ice cover? If the land slowly rose when freed from ice and carried the bones of whales to the summits of hills, why did the neighbouring land subside miles deep, as the undersea canyons indicate?

Daly concluded: "The Pleistocene history of North America holds ten major mysteries for every one that has already been solved."[7]

[6] Ibid., p. 362.
[7] Daly, *The Changing World of the Ice Age* (1934), p. 111.

Chapter V

TIDAL WAVE

Fissures in the Rocks

Joseph Prestwich, professor of geology at Oxford (1874–88) and acknowledged authority on the Quaternary (Glacial and Recent) Age in England, was struck by numerous phenomena, all of which led him to the belief that "the south of England had been submerged to the depth of not less than about 1000 feet between the Glacial—or Post-Glacial—and the recent or Neolithic [Late Stone] periods."[1] In a spasmodic movement of the terrain, the coast and the land masses of southern England were submerged to such a depth that points 1000 feet high were below sea level.[2]

A most striking phenomenon among those observed by Prestwich was in the fissures in the rocks. In the neighbourhood of Plymouth on the Channel, clefts of various widths in limestone formations are filled with rock fragments, angular and sharp, and with bones of animals—mammoth, hippopotamus, rhinoceros, horse, polar bear, bison. The bones are "broken into innumerable fragments. No skeleton is found entire. The separate bones, in fact, have been dispersed in the most irregular manner, and without any bearing to their relative position in the skeleton. Neither do they show wear, nor have they been

[1] Joseph Prestwich, "The Raised Beaches and 'Head' or Rubble-drift of the South of England," *Quarterly Journal of the Geological Society*, XLVIII (1892), 319–37; Prestwich, "On the Evidences of a Submergence of Western Europe and of the Mediterranean Coasts at the Close of the Glacial or So-called Post-Glacial Period, and Immediately Preceding the Neolithic or Recent Period," *Philosophical Transactions of the Royal Society of London, 1893,* Series A (1894), pp. 904ff.
[2] Ibid., p. 906.

gnawed by beasts of prey, though they occur with the bones of hyaena, wolf, bear and lion."[3] In other places in Devonshire and also in Pembrokeshire in Wales, ossiferous breccia or conglomerates of broken bones and stones in fissures in limestone consist of angular rock fragments and "broken and splintered" bones with sharp fractured edges in a "fresh state," and in "splendid condition," showing no traces of gnawing.[4]

If the crevices were pitfalls into which the animals fell alive, then some of the skeletons would have been preserved entire. But this is "never the case." "Again if left for a time exposed in the fissures, the bones would be variously weathered, which they are not. Nor would the mere fall have been sufficient to have caused the extensive breakage the bones have undergone: these, I consider, are fatal objections to this explanation, and none other has since been offered," wrote Prestwich.[5]

Fissures in the rocks, not only in England and Wales, but all over western Europe, are choked with bones of animals, some of extinct races, others, though of the same age, of races still surviving. Osseous breccia in the valleys around Paris have been described, as well as fissures in the rocks on the tops of isolated hills in central France. They contain remnants of mammoth, woolly rhinoceros, and other animals. These hills are often of considerable height. "One very striking example"[6] is found near Semur in Burgundy: a hill—Mont Genay—1430 feet high is capped by a breccia containing remains of mammoth, reindeer, horse, and other animals.

In the rock on the summit of Mont de Sautenay—a flat-topped hill near Chalon-sur-Saône between Dijon and Lyons—there is a fissure filled with animal bones. "Why should so many wolves, bears, horses, and oxen have ascended a hill isolated on all sides?" asked Albert Gaudry, professor at the Jardin des Plantes. According to him, the bones in this cleft are mostly broken and splintered into innumerable sharp fragments and are "evidently not those of animals devoured by beasts of prey; nor have they been broken by man. Nevertheless, the remains of wolf were particularly abundant, together with those of cave

[3] Prestwich, *On Certain Phenomena Belonging to the Close of the Last Geological Period and on Their Bearing upon the Tradition of the Flood* (London: Macmillan and Co., 1895), pp. 25–26.

[4] Prestwich, *Quarterly Journal of the Geological Society*, XLVIII, 336.

[5] Prestwich, *On Certain Phenomena*, p. 30.   [6] Ibid., p. 36.

lion, bear, rhinoceros, horse, ox, and deer. It is not possible to suppose that animals of such different natures, and of such different habitats, would in life ever have been together."[7] Yet the state of preservation of the bones indicates that the animals —all of them—perished in the same period of time. Prestwich thought that the animal bones, "now associated in the fissure on the summit of the hill," were found in common heaps because, "we may suppose, all these animals had fled [there] to escape the rising waters."[8]

On the Mediterranean coast of France there are numerous clefts in the rocks crammed to overflowing with animal bones. Marcel de Serres wrote in his survey of the Montagne de Pédémar in the Department of Gard: "It is within this limited area that the strange phenomenon has happened of the accumulation of a large quantity of bones of diverse animals in hollows or fissures."[9] De Serres found the bones all broken into fragments, but neither gnawed nor rolled. No coprolites (hardened animal faeces) were found, indicating that the dead beasts had not lived in these hollows or fissures.

The Rock of Gibraltar is intersected by numerous crevices filled with bones. The bones are broken and splintered. "The remains of panther, lynx, caffir-cat, hyaena, wolf, bear, rhinoceros, horse, wild boar, red deer, fallow deer, ibex, ox, hare, rabbit, have been found in these ossiferous fissures. The bones are most likely broken into thousands of fragments—none are worn or rolled, nor any of them gnawed, though so many carnivores then lived on the rock," says Prestwich,[10] adding: "A great and common danger, such as a great flood, alone could have driven together the animals of the plains and of the crags and caves."[11]

The Rock is extensively faulted and fissured. Beaches high on Gibraltar show that the expression that makes of this rock the symbol of immovability is unfounded. These beaches indicate that at some time the waters of the sea lapped the Rock at the

[7] Ibid., pp. 37–38.   [8] Ibid., p. 38.
[9] Marcel de Serres, "Note sur de nouvelles brèches osseuses découvertes sur la montagne de Pédémar dans les environs de Saint-Hippolyte-du-Forte (Gard)," Bulletin du Société Géologique de France, 2e. Série, XV (1858), p. 233.
[10] Prestwich, On Certain Phenomena, p. 47; Idem, Philosophical Transactions of the Royal Society, 1893. p. 935.
[11] Prestwich, On Certain Phenomena, p. 48.

6oo-foot mark; the Rock now rises over 1370 feet above the sea. It was therefore, "in Quaternary times [or the age of man], an island not more than about 800 feet, or less high, which rose by successive stages to its present height. It is more than probable, however, that at some time before it settled at that level, the whole of the area was upheaved to such an extent that a land passage was formed to the African coast. . . ."[12]

A human molar and some flints worked by Paleolithic (old stone) man, as well as broken pieces of pottery of Neolithic (recent or polished stone) man, were discovered among the animal bones in some of the crevices of the Rock.[13]

On Corsica, Sardinia, and Sicily, as on the continent of Europe and the British Isles, the broken bones of animals choke the fissures in the rocks. The hills around Palermo in Sicily disclosed an "extraordinary quantity of bones of hippopotami—in complete hecatombs." "Twenty tons of these bones were shipped from around the one cave of San Ciro, near Palermo, within the first six months of exploiting them, and they were so fresh that they were sent to Marseilles to furnish animal charcoal for use in the sugar factories. How could this bone breccia have been accumulated? No predaceous animals could have brought together and left such a collection of bones."[14] No teeth marks of hyena or of any other animals are found in this osseous mass. Did the animals come there to die as old age approached? "The bones are those of animals of all ages down to the foetus, nor do they show traces of weathering or exposure."[15]

"The extremely fresh condition of the bones, proved by the retention of so large a proportion of animal matter," shows that "the event was, geologically, comparatively recent"; and the "fact that animals of all ages were involved in the catastrophe" shows it "to have been sudden." Prestwich was of the opinion that, together with central Europe and England, the Mediterranean islands, Corsica, Sardinia, and Sicily, had been submerged. "The animals in the plain of Palermo naturally retreated, as the waters advanced, deeper into the amphitheatre of hills until they found themselves embayed . . . the animals must have thronged together in vast multitudes, crushing into the more accessible caves, and swarming over the ground at their entrance, until overtaken by the waters and destroyed.

[12] Ibid., p. 46.　[13] Ibid., p. 48.　[14] Ibid., p. 50.　[15] Ibid., p. 51.

. . . Rocky debris and large blocks from the sides of the hills were hurled down by the current of water, crushing and smashing the bones."[16]

Prestwich, who subscribed to the Ice Age theory and is regarded as the foremost authority on the geology of the Ice Age in England, was compelled to construct a theory of "submergence of Western Europe and of the Mediterranean coasts at the close of the Glacial or so-called Post-Glacial Period, and immediately preceding the Neolithic or Recent Period," which quotation was the title of a paper read by him before the Royal Society of London. It was published in the Society's *Philosophical Transactions*. It became clear to Prestwich that it was "impossible to account for the specific geological phenomena . . . by any agency of which our time has offered us experience."[17] "The agency, whatever it was, must have acted with sufficient violence to smash the bones."[18] "Nor could this have been the work of a long time, for the entombed bones, though much broken, are singularly fresh."[19] "Certain communities of early man must have suffered in the general catastrophe."[20]

The Rock of Gibraltar rose to close the strait, then sank down part way; the coast of England and even hills 1000 feet high were submerged; the island of Sicily was inundated, as were elevations in the interior of France. Everywhere the evidence betokens a catastrophe that occurred in not too remote times and engulfed an area of at least continental dimensions. Great avalanches of water loaded with stones were hurled on the land, shattering massifs, and searching out the fissures among the rocks, rushed through them, breaking and smashing every animal in their way.

In Prestwich's opinion the cause of the catastrophe was the sinking of the continent and its subsequent elevation, which was sudden, and during which water from the heights broke upon lower levels, bringing chaos and destruction. Prestwich suspected that the area involved must have been much larger than that discussed in his works. He gave no reason for such submergence and emergence. The catastrophe occurred when England was entering the age of polished stone, or, possibly, when the centres of ancient civilization were in the Bronze Age.

[16] Ibid., pp. 51–52. [17] Ibid., p. vi. [18] Ibid., p. 67. [19] Ibid., p. 7.
[20] Ibid., p. 74.

In a later section of this book are presented archaeological evidences of vast catastrophes that more than once shattered every city and settlement of the ancient world: Crete, Asia Minor, the Caucasus, Mesopotamia, Iran, Syria, Palestine, Cyprus, and Egypt were simultaneously and repeatedly laid waste. These catastrophes occurred when Egypt was in the Bronze Age and when Europe was entering the Neolithic Age.

The Norfolk Forest-Bed

As an area is investigated, more problems are raised than are solved. Britain is the land of great geologists, the founders and leaders of that science, and the soil of Britain has been explored more than any other soil on the five continents or in the seven seas. Examination of Britain's record of the Ice Age levels discloses a "complex interbedding of drift sheets derived from different sources." "When we add the additional complications imposed by thin drifts, scanty interglacial deposits, and the frequent presence in fossil-bearing beds of secondary [displaced] fossils derived from the reworking of older horizons, we get a truly difficult over-all problem. . . . All in all, British glacial stratigraphic research has encountered exceptional difficulties," writes R. F. Flint, professor of geology at Yale University.[1]

In Cromer, Norfolk, close to the North Sea coast, and in other places on the British Isles, "forest-beds" have been found. The name derives from the presence of a great number of stumps of trees once supposed to have rooted and grown where they are now found. Many of the stumps are in upright positions, and their roots are often interlocked. Today these forests are recognized as having drifted: the roots do not end in small fibres, but are broken off, in most cases one to three feet from the trunk.

Bones of sixty species of mammals, besides birds, frogs, and snakes, were found in the forest-bed of Norfolk. Among the mammals were the sabre-toothed tiger, huge bear (*Ursus horribilis*), mammoth, straight-tusked elephant, hippopotamus, rhinoceros, bison, and modern horse (*Equus caballus*). Two exclusively northern species—glutton and musk-ox—were found

[1] Flint, *Glacial Geology and the Pleistocene Epoch*, p. 377.

51

among animals from temperate and tropical latitudes. Of the thirty species of large land animals of the forest-bed, only six still exist in any part of the world—all the others are extinct—and only three are presently native to the British Isles.[2]

Remains of sixty-eight species of plants were obtained from the Norfolk forest-bed; they indicate "a climate and geographical conditions very similar to those of Norfolk at the present day."[3] In view of the sensitivity of plants to thermal conditions, the conclusion might well be drawn that the climate at the time the forest-bed was deposited was not different from the present, which conclusion the fauna, comprising southern as well as northern animals, contradicts.

The abundance of animals of so many different species on an island the size of Great Britain cause speculation that at one time it must have been part of a continent and that the Strait of Dover was not then opened. It was further conjectured that the Rhine flowed on to the north across the area at present occupied by the sea—the Thames being one of it tributaries—and that the estuary of the Rhine was for some time at Cromer; that the trees were carried there by the Rhine; that they grew on the banks of the river, and the water washed out their roots and the falling trunks were carried away and deposited as the forest-bed. "It is necessary to point out, however, that the opening of the Straits of Dover is a geological revolution of considerable magnitude, such as one might well hesitate to ascribe to the comparatively short period embraced by glacial and post-glacial time."[4]

Immediately above the forest-bed there is a fresh-water deposit with arctic plants—arctic willow and dwarf birch—and land shells. It is "a remarkable change from the climatic conditions of the Forest-bed below. . . . [It] is such as to indicate a lowering of temperature of about 20°."[5]

On top of the arctic fresh-water plants and shells is a marine bed. *Astarte borealis* and other mollusc shells are found "in the position of life, with both valves united." These species "are arctic, but, as the bed seems in other places to contain *Ostrea edulis* [a mollusc], which requires a temperate sea, the evidence is conflicting as to the climate."[6]

[2] W. B. Wright, *The Quaternary Ice Age* (1937), p. 110.
[3] Ibid. [4] Ibid., p. 111. [5] Ibid. [6] Ibid.

What could have brought, together or in quick succession, all these animals and plants, from the tundra of the Arctic Circle and from the jungle of the tropics, from lush oak forest and from desert, from lands of many latitudes and altitudes, from fresh-water lakes and rivers, and from the salt seas of the north and the south? The shells with closed valves furnish evidence that the molluscs did not die a natural death but were buried alive.

It would appear that this agglomeration was brought together by a moving force that rushed overland, left in its wake marine sand and deep-water creatures, swept animals and trees from the south to the north, and then, turning from the polar regions back toward the warm regions, mixed its burden of arctic plants and animals in the same sediment where it had left those from the south. Animals and plants of land and sea from various parts of the world were thrown together, one group upon another, by some elemental force that could not have been an overflowing river. Also bones of animals already extinct in earlier epochs were carried out of their beds and thrown into the jumble.

The finding of warm-climate animals and plants in polar regions, coral and palms in the Arctic Circle, presents these alternatives: either these animals and plants lived there at some time in the past or they were brought there by tidal waves. In some cases the first is true, as where stumps of trees (palms) are found *in situ*. In other cases the second is true, as where, in one and the same deposit, animals and plants from sea and land, from south and north, are found in a medley. But in both cases one thing is apparent: such changes could not have occurred unless the terrestrial globe veered from its path, either because of a disturbance in the speed of rotation or because of a shift in the astronomical or geographical position of the terrestrial axis.

In many cases it can be shown that southern plants grew in the north; either the geographical position of the pole and the latitudes or the inclination of the axis must have changed since then. In many other cases it can be shown that a marine irruption threw into one deposit living creatures from the tropics and from the Arctic; the change must have been sudden, instantaneous. We have both kinds of cases. Consequently there must have been changes in the position of the axis, and they must have been sudden.

Cumberland Cavern

In 1912 near Cumberland, Maryland, workmen cutting the way for a railroad with dynamite and steam shovel came upon a cavern or a closed fissure with "a peculiar assemblage of animals. Many of the species are comparable to forms now living in the vicinity of the cave; but others are distinctly northern or Boreal in their affinities, and some are related to species peculiar to the southern, or Lower Austral, region." Thus wrote J. W. Gidley and C. L. Gazin of the United States National Museum.[1]

A crocodilid and a tapir are representative of southern climate; a wolverine and a lemming "are distinctly northern." It seems "highly improbable" that they co-existed in one place; the usual assumption was made that the cave received the animal remains in a glacial and an interglacial period. However, the scientist who explored the cavern for the Smithsonian Institution as soon as it was discovered and who returned there in the following years for closer investigation, J. W. Gidley, contended that the animals were contemporaneous: the position of the bones excluded any other explanation. "This strange assemblage of fossil remains occurs hopelessly intermingled. . . ."[2]

The bones of the Cumberland cavern were "for the most part much broken, yet show no sign of being water worn."[3] This would signify that the bones were not carried for any length of time by a stream; however, it is quite possible that the animals were dashed against the rocks by an avalanche of water that carried them far off, broke their bones inside their bodies—thus the bones are not water-worn—and there smashed togther all kinds of animals; then gravel and rocks enclosed them.

So also it happened that animals of northern regions—wolverine and lemming, the long-tailed shrew, mink, red squirrel,

[1] J. W. Gidley and C. L. Gazin, *The Pleistocene Vertebrate Fauna from Cumberland Cave, Maryland*, U.S. National Museum Bulletin 171 (1938).

[2] Gidley in *Explorations and Field-work of the Smithsonian Institution for the Year 1913* (Washington, 1914); *Annual Report of the Smithsonian Institution for 1918*, pp. 281–87.

[3] *Explorations and Field-work of the Smithsonian Institution for the Year 1913*, pp. 94–95.

muskrat, porcupine, hare, and elk—were heaped together with animals "suggesting warmer climatic conditions"—peccary, crocodilid, and tapir. Animals that now live on the western coast of America—coyote, badger, and puma-like cat—are in this assemblage. Animals that live in areas of plentiful water supply—beaver and muskrat and mink—are found in the Cumberland cavern jumbled together with animals of arid regions—coyote and badger—and those of wooded regions together with animals of open terrain, like the horse and the hare. This is truly "a peculiar assemblage of animals." Extinct animals are found there intermingled with extant forms. Death came to all of them at the same time. Any theory that attempts to explain the presence of animal bones from various climates in one and the same locality by a sequence of glacial and interglacial periods must stumble on the bones of the Cumberland cavern.

In Northern China

In the village of Choukoutien, near Peiping (Peking) in northern China, in caverns and in fissures in rocks, a great mass of animal bones was found. "The most astonishing fact was the discovery of this unimaginable wealth of bones of fossil animals" (Weidenreich). These rich ossiferous deposits occur in association with human skeletal remains.

"As Weidenreich began his studies, other amazing, nearly unexplainable features appeared." The fractured bones of seven human individuals were found there. 'A European, a Melanesian, and an Eskimo type lying dead in one close-knit group in a cave, on a Chinese hillside! Weidenreich marvelled."[1] It was assumed that the seven inhabitants of the narrow fissure were murdered because their skulls and bones are fractured. It is possible that these several types of man came together in Choukoutien, since the migrations of ancient man were on a greater scale than is generally thought.

But the finders of the conglomerates of bones were perplexed also by the animal remains; the bones belonged to animals of the tundras, or a cold-wet climate; of steppes and prairies, or dry climate; and of jungles, or warm-moist climate, "in a strange

[1] R. Moore, *Man, Time, and Fossils* (1953), pp. 274–75.

mixture." Mammoths and buffaloes and ostriches and arctic animals left their teeth, horns, claws, and bones in one great melange, and though we have met very similar situations in various places in other parts of the world, the geologists of China regarded their find as enigmatic.

"No conclusive evidence can be derived from this faunal assemblage as regards the prevailing temperature at the time when it lived," says J. S. Lee in his *Geology of China*.[2] Some animals point "to a rather severe climate," other animals to "a warm climate." "It is almost inconceivable" that animals of such various habitats should live together. "And yet their remains are found side by side."

It is asserted that since before the age of man—since late Tertiary times and through the time of the Great Ice Age in Europe and America—northern China experienced "progressive desiccation interrupted by pluvial intervals."[3] Arid conditions prevailed over northern China and "the general absence of ice-sculptured features" led the naturalists to the conclusion that in northern China, as in northern Siberia, there were no glacial conditions and no formation of ice cover. "On the other hand, certain obscure facts not in agreement with the foregoing interpretation are accumulating throughout the country."[4] Erratic blocks and striated boulders are found in the valleys and on the hills.

But if there was no ice cover in northern China or in Siberia to the north, what was it that carried the bones of animals into fissures in the rocks? And what striated the rocks and transported boulders far from the source of their origin and high on to hills?

At the same time convincing evidence was brought forth that "the mountain ranges in western China have been elevated since the Glacial Age."[5]

At Tientsin marine sands and clays with the shells of sea molluscs have been found exposed on the surface of the ground. Borings made in the same location "showed the presence of sand and clays with fresh-water shells down to a depth of more than 507 feet below the marine layer which is exposed on the surface."[6] Thus signs of both recent elevation and submergence are present.

[2] J. S. Lee, *The Geology of China* (London, 1939), p. 370.
[3] Ibid., p. 371. [4] Ibid. [5] Ibid., p. 207. [6] Ibid., p. 206

Was not the irrupted sea the agent that threw together the animals of various latitudes and carried rocks of foreign origin to the tops of hills? Did not the mountains that sprang up in the age of man rise in the upheaval that also moved the seas out of their borders?

Were not animals of various habitats swept into fissures—human beings with them—when mountains rose, seas irrupted, rock debris was carried toward summits, and climate changed?

The fossils of Choukoutien are found imbedded in a reddish loam, a mixture of clay and sand, the deposition of which belongs to the same stage as the fossils; this reddish loam occurs extensively all over northern China. Teilhard and Young concluded that the observed coloration "can neither be a quality inherited from the original material of which the loams are composed, nor a condition brought about by slow chemical processes long after their formation." The coloration of this widespread formation being of some extraneous and unexplained origin, the only definite statement concerning it is that some violent change of climate, in itself not the cause of the change of colour, occurred "immediately before the deposition of red loams—or soon after the deposition."[7]

Similar observations were made in other parts of the world. Drift, the displacement of which is attributed to the ice cover, is often found tinted a reddish colour. R. T. Chamberlin, looking for the origin of this hue, offered the hypothesis that "granite pebbles were decomposed, the liberated iron staining the drift reddish."[8]

H. Pettersson, of the Oceanographic Institute at Göteborg, on examining red clay from the bottom of the Pacific, found that the abysmal clay contains layers of ash and a high content of nickel, almost completely absent in the water.[9] Pettersson, whose work will be described on a later page, attributed the origin of nickel and iron in the clay to prodigious showers of meteorites; the lavas of the oceanic bedrock he recognized as of recent origin."[10]

[7] J. S. Lee, *The Geology of China*, pp. 202, 368, 371.
[8] Chamberlin in *Man and Science*, ed. Moulton, p. 92.
[9] H. Pettersson, "Chronology of the Deep Ocean Bed," *Tellus* (*Quarterly Journal of Geophysics*), I (1949).
[10] See the section, "The Floor of the Seas."

All this points to a great shower of ferruginous dust at a recent geological date, when the red clays of the Pacific, the drift of the Western Hemisphere, and the loam of China were deposited, and when the climate also changed.

The Asphalt Pit of La Brea

At Rancho La Brea, once on the western outskirts of Los Angeles, and at present in the immediate neighbourhood of the luxurious shopping centre of that city, bones of extinct animals and of still living species are found in abundance in asphalt mixed with clay and sand. In 1875 some fossil remains of this bituminous deposit were described for the first time. By then thousands of tons of asphalt had already been removed and shipped to San Franscisco for roofing and paving.[1]

Beds of petroleum shale (rock of laminated structure formed by the consolidation of clay), ascribed to the Tertiary Age, having in many places a thickness of about two thousand feet, extend from Cape Mendocino in northern California to Los Angeles and beyond, a distance of over four hundred and fifty miles. The asphalt beds of Rancho La Brea are an outcrop of this large bituminous formation.

Since 1906 the University of California has been collecting the fossils of Rancho La Brea, "a most remarkable mass of skeletal material." When found, these fossils were regarded as representing the fauna of the late Tertiary (Pliocene) or early Pleistocene (Ice Age). The Pleistocene strata, fifty to one hundred feet thick, overlie the Tertiary formations in which the main oil-bearing beds are found. The deposit containing the fossils consists of alluvium, clay, coarse sand, gravel, and asphalt.

Most spectacular among the animals found at Rancho La Brea is the sabre-toothed tiger (*Smilodon*), previously unknown elsewhere in the New or Old World, but found since then in other places too. The canine teeth of this animal, over ten inches long, projected from his mouth like two curved knives. With this weapon the tiger tore the flesh of his prey.

[1] Cf. J. C. Merriam, "The Fauna of Rancho La Brea," *Memoirs of the University of California*, I, No. 2 (1911).

The animal remains are crowded together in the asphalt pit in an unbelievable agglomeration. In the first excavation carried on by the University of California "a bed of bones was encountered in which the number of sabre-tooth and wolf skulls together averaged twenty per cubic yard."[2] No fewer than seven hundred skulls of the sabre-toothed tiger have been recovered.[3]

Among other animals unearthed in this pit were bison, horses, camels, sloths, mammoths, mastodons, and also birds, including peacocks.

In the time following the discovery of America this region of the coast was rather sparsely populated with aniamls; early immigrants found only "semi-starved coyotes and rattle-snakes."[4] But when Rancho La Brea received its skeletons "there lived an amazing assemblage of animals in Western America."[5]

To explain the presence of these bones in the asphalt, the theory was offered that the animals became entrapped in the tar, sank in it, and were embedded in asphalt when the tar hardened. However, the large number of animals that filled this asphalt bed to overflowing is baffling. Moreover, the fact that the vast majority of them are carnivorous, whereas in any fauna the majority of animals would be herbivorous—otherwise the carnivores would have no victims for their daily food—requires explanation. So it was assumed that some animal, caught in the tar, cried out, thus attracting more of its kind, and these were trapped, too, and at their cries carnivorous animals came, followed by more and more.

This explanation might be valid if the state of the bones did not testify that the ensnarement of the animals by the tar happened under violent circumstances. Oil from which the volatile elements have evaporated leaves asphalt, tar, and other bitumens. "As the greater number of the animals in the Rancho La Brea beds have been entrapped in the tar, it is to be presumed that in a large percentage of cases the major portion of the skeleton has been preserved. Contrary to expectations, connected skeletons are not common."[6] The bones are "splendidly"

[2] Ibid.   [3] R. S. Lull, *Fossils* (1931), p. 28.
[4] George McCready Price, *The New Geology* (1923), p. 579.
[5] Lull, *Fossils*, p. 27.
[6] Merriam. *Memoirs of the University of California*, I, No. 2.

preserved[7] in the asphalt, but they are "broken, mashed, contorted, and mixed in a most heterogeneous mass, such as could never have resulted from the chance trapping and burial of a few stragglers."[8]

Were not the herds of frightened animals found at La Brea engulfed in a catstrophe? Could it be that at this particular spot large herds of wild beasts, mostly carnivorous, were overwhelmed by falling gravel, tempests, tides, and raining bitumen?[9] Similar finds in asphalt have been unearthed in two other places in California, at Carpinteria and McKittrick; the depositions were made under comparable circumstances. The plants of the Carpinteria tar pits were found, with one exception, to have been "members of the Recent flora," or of the flora now living 200 miles to the north.[10]

Separate bones of a human skeleton were also discovered in the asphalt of La Brea. The skull belonged to an Indian of the Ice Age, it is assumed. However, it does not show any deviation from the normal skulls of Indians.

The human bones were found in the asphalt under the bones of a vulture of an extinct species. These finds suggest that the time when the human body was buried preceded the extinction of that species of vulture or at least coincided with it; in a turmoil of elements the vulture met its death, as did possibly the rest of its kind, with the sabre-toothed tiger and many other species and genera.

Agate Spring Quarry

In Sioux County, Nebraska, on the south side of the Niobrara River, in Agate Spring Quarry, is a fossil-bearing deposit up to twenty inches thick. The state of the bones indicate a long and violent transportation before they reached their final resting place. "The fossils are in such remarkable profusion in places as to form a veritable pavement of interlacing bones, very few

[7] Lull, *Fossils*, p. 28. [8] Price, *The New Geology*, p. 579.
[9] C. E. Brasseur, *Histoire des nations civilisées du Mexique*, (1857–59), I, 55; *Popul-Vuh, le livre sacre*, ed. Brasseur (1861), p. 25.
[10] R. W. Chaney and H. L. Mason, "A Pleistocene Flora from the Asphalt Deposits at Carpinteria, California," in *Studies of the Pleistocene Paleobotany of California* (Carnegie Institution, 1934).

of which are in their natural articulation with one another," says R. S. Lull, director of the Peabody Museum at Yale, in his book on fossils.[1]

The profusion of bones in Agate Spring Quarry may be judged by a single block now in the American Museum of Natural History in New York. This block contains about 100 bones to the square foot. There is no way of explaining such an aggregation of fossils as a natural death retreat of animals of various genera.

The animals found there were mammals. The most numerous was the small twin-horned rhinoceros (*Diceratherium*). There was another extinct animal (*Moropus*) with a head not unlike that of a horse but with heavy legs and claws like those of a carnivorous animal; and bones of a giant swine that stood six feet high (*Dinohyus hollandi*) were also unearthed.

The Carnegie Museum, which likewise excavated in Agate Spring Quarry, in a space of 1350 square feet found 164,000 bones or about 820 skeletons. A mammal skeleton averages 200 bones. This area represents only one twentieth of the fossil bed in the quarry, suggesting to Lull that the entire area would yield about 16,400 skeletons of the twin-horned rhinoceros, 500 skeletons of the clawed horse, and 100 skeletons of the giant swine.

A few miles to the east, in another quarry, were found skeletons of an animal which, because of its similarity to two extant species, is called a gazelle camel (*Stenomylus*). A herd of these animals was destroyed in a disaster. As in Agate Spring Quarry, the fossil bones were deposited in sand transported by water. The transportation was in a violent cataract of water, sand, and gravel, that left marks on the bones.

Tens of thousands of animals were carried over an unknown distance, then smashed into a common grave. The catastrophe was most probably ubiquitous, for these animals—the small twin-horned rhinoceros, clawed horse, giant swine, and gazelle camel—did not survive, but became extinct. There is nothing in their skeletons to warrant regarding them as degenerate and doomed to extinction. And the very circumstances in which they are found bespeak a violent death at the hands of the elements, not slow extinction in a process of evolution.

In many other places of the world similar finds have been

[1] Lull, *Fossils*, p. 34.

made, and in one of the sections to follow we shall discuss the famous bone quarry of Siwalik. In the United States, Big Bone Lick, Kentucky, twenty miles south of Cincinnati, contained the bones of one hundred mastodons, besides many other extinct animals. President Jefferson gathered there his famous collection of fossils. In San Pedro Valley, California, skeletons of the mastodon are found standing upright, in the posture in which they died, mired in gravel, ash, and sand. Fossils found in John Day Basin, Oregon, and the glacial Lake Florissant, Colorado, are embedded in volcanic ash. In the Southern states fossil bones are quarried for the commercial exploitation of phosphates.

In Switzerland a conglomerate of bones of animals that belong to different climates and habitats was found in Kesslerloch near Thayngen: Alpine types are there in one "Tiergemisch" with animals of the steppe and of the forest fauna.[2] In Germany a gravel pit at Neuköln (formerly Rixdorf), a suburb of Berlin, disclosed two faunas: mammoth, musk ox, reindeer, and arctic fox "suggest a boreal climate"; lion, hyena, bison, ox, and two species of elephant "suggest varying degrees of warmer climate." The faunas were interpreted as belonging to two periods— glacial and interglacial—but the bones were found all together. "It seems probable that the relations are more complicated than has been realized."[3] There has not yet been found "a satisfactory climatic interpretation."

Great multitudes of animals that filled prairies and forests, water and air, forms, fragile or sturdy, with an urge to live and multiply, were more than once suddenly called upon to write their names in the register of extinction.

[2] Heierli, "Das Kesslerloch bei Thayngen," *Neue Denkschriften der Schweizerischen Naturforschenden Gesellschaft*, Vol XLIII (1907); H. Broekmann-Jerosch in *Die Veränderungen des Klimas*, publ, by the XI-th International Geological Congress (1910).

[3] Flint, *Glacial Geology*, p. 329.

Chapter VI

MOUNTAINS AND RIFTS

Mountain Thrusts in the Alps and Elsewhere

The age of a rock formation is ascertained with the help of the fossils it contains. To the surprise of many scientists, it was found that mountains have travelled, since older formations have been pushed over on top of younger ones.

Chief Mountain in Montana is a massif standing several thousand feet above the Great Plains. It "has been thrust bodily upon the much younger strata of the Great Plains, and then driven over them eastward, for a distance of at least eight miles. Indeed, the thrust may have been several times eight miles," writes Daly.[1]

"By similar thrusting, the whole Rocky Mountain Front, for hundreds of miles, has been pushed up and then out, many miles over the plains."[2]

Such titanic displacements of mountains have been found in many places on the earth. The displacement of the Alps is especially extensive.

"During the building of the Alps gigantic slabs of rock, thousands of feet thick, hundreds of miles long, and tens of miles wide, were thrust up and then over, relatively to the rocks beneath. The direction of the relative overthrusting movement was from Africa toward the main mass of Europe on the north. The visible rocks of the northern Alps of Switzerland have thus been shoved northward distances of the order of 100 miles. In a sense the Alps used to be on the present site of northern Italy."[3]

[1] Daly, *Our Mobile Earth*, pp. 228–29.   [2] Ibid., p. 231.
[3] Ibid., pp. 232–33.

63

Mount Blanc was moved from its place and the Matterhorn was overturned.

Those portions of the Alps that surround the valley of the Linth, in the canton of Glarus in Switzerland, have lower parts of Tertiary formations or of the age of mammals; their upper parts are Permian (preceding the age of reptiles) and Jurassic (of the age of reptiles). This impels to one of two conclusions: either the division of rocks into sequences based on the fossils they contain is fallacious, or the old mountains were moved bodily and set on the shoulders of more recent formations. The latter conclusion is chosen; and if De Saussure's notion of the sea sweeping over the Alps appeared fantastic, the idea of mountains travelling considerable distances must sound even more fantastic, unless we know of a physical cause that could have brought it about. But even the very cause of mountain building itself is obscure.

"The problem of mountain-making is a vexing one: Many of them [mountains] are composed of tangentially compressed and overthrust rocks that indicate scores of miles of circumferential shortening in the Earth's crust. Radial shrinkage is woefully inadequate to cause the observed amount of horizontal compression. Therein lies the real perplexity of the problem of mountain-making. Geologists have not yet found a satisfactory escape from this dilemma," says F. K. Mather of Harvard University.[4]

The origin of the mountains is not explained; and still less is their thrust or shift across valleys and over other mountains. The Alps were shoved a hundred miles to the north. Chief Mountain in Montana travelled across the plains and climbed the slope of another mountain and settled on top of it. ". . . All of the Glacier National Park in Montana and all the Rocky Mountain area up to the Yellowhead Pass in Alberta" moved for many miles.[5] The mountains of western Scotland shifted from their places. The entire length of the Norwegian mountains showed a similar overthrust. What could have caused these mountains to travel across valleys and uphill with their masses

[4] F. K. Mather, reviewing G. Gamow, *Biography of the Earth*, in *Science*, January 16, 1942.
[5] George McCready Price, *Common-sense Geology*, p. 120, Idem, "The Fossils as Age-makers in Geology," *Princeton Theological Review*, Vol. XX, No. 4, October 1922.

of granite weighing billions of tons? No force acting from inside the earth, pulling inward or pushing outward, could have created these overthrusts. Only twisting could have produced them. It could hardly have occurred if the rotation and revolution of our planet had never been disturbed.

In the Alps, caverns with human artifacts of stone and bone dating from the Pleistocene (Ice Age) have been found at remarkably high altitudes. During the Ice Age the slopes and valleys of the Alps, more than other parts of the continent, must have been covered by glaciers; today in central Europe there are great glaciers only in the Alps. The presence of men at high altitudes during the Pleistocene or Paleolithic (rude stone) Age seems baffling.

The cavern of Wildkirchli, near the top of Ebenalp, is 4900 feet above sea level. It was occupied by man sometime during the Pleistocene. "Even more remarkable, in respect to altitude, is the cavern of Drachenloch at a height of 2445 metres [8028 feet]," near the top of Drachenberg, south of Ragaz. This is a steep, snow-covered massif. "Both of these stations are in the very heart of the Alpine field of glaciation."[6]

A continental ice sheet thousands of feet thick filled the entire valley between the Alps and the Jura, where now Lake Geneva lies, to the height of the erratic boulders torn from the Alps and placed on the Jura Mountains. In the same geological epoch, between two advances of the ice cover, during an interglacial intermission, human beings must have occupied caverns 8000 feet above sea level. No satisfactory explanation for such location of Stone Age man has ever been offered.

Could it be that the mountains rose as late as in the age of man and carried up with them the caverns of early man? In recent years evidence has grown rapidly to show, in contrast to previous opinion, that the Alps and other mountains rose and attained their present heights, and also travelled long distances, in the age of man.

"Mountain uplifts amounting to many thousands of feet have occurred within the Pleistocene epoch [Ice Age] itself." This occurred with "the Cordilleran mountain system in both North

[6] G. G. MacCurdy, *Human Origins* (1924), I, 77.

and South America, the Alps-Caucasus-Central Asian system, and many others. . . ."[7]

The fact of the late upthrust of the major ridges of the world created, when recognized, great perplexity among geologists who, under the weight of much evidence, were forced to this view. The revision of the concepts is not always radical enough. Not only in the age of man, but in the age of *historical* man, mountains were thrust up, valleys were torn out, lakes were dragged uphill and emptied. Helmut Gams and Rolf Nordhagen brought together very extensive material concerning the Bavarian Alps and the Tyrol, or Eastern Alps. We shall deal with this material in Chapter XI, "Klimasturz."

"The great mountain chains challenge credulity by their extreme youth," wrote the explorer Bailey Willis about Asian mountains.[8]

The Himalayas

The Himalayas, highest mountains in the world, rise like a thousand-mile-long wall north of India. This mountain wall stretches from Kashmir in the west to and beyond Bhutan in the east, with many of its peaks towering over 20,000 feet, and Mount Everest reaching 29,000 feet, or over five miles. The summits of these lofty massifs are capped by eternal snow in those regions of the heavens where eagles do not fly nor any other bird of the sky.

Scientists of the nineteenth century were dismayed to find that, as high as they climbed, the rocks of the massifs yielded skeletons of marine animals, fish that swim in the ocean, and shells of molluscs. This was evidence that the Himalayas had risen from beneath the sea. At some time in the past azure waters of the ocean streamed over Mount Everest, carrying fish, crabs, and molluscs, and marine animals looked down to where now we look up and where man, after many unsuccessful efforts, has until now succeeded only once in putting his feet. Until recently it was assumed that the Himalayas rose from the bottom of the sea to their present height tens or perhaps hundreds of millions of years ago. Such a long period of time, so

[7] Flint, *Glacial Geology and the Pleistocene Epoch*, pp. 9–10.
[8] B. Willis, *Research in Asia*, II, 24.

long ago, was enough even for the Himalayas to have risen to their present height. Do we not, when we tell young listeners a story about giants and monsters, begin with: "Once upon a time, long, long ago . . ."? And the giants are no longer threatening and the monsters are no longer real.

According to the general geological scheme, five hundred million years ago the first forms of life appeared on earth; two hundred million years ago life developed into reptilian forms that dominated the scene, achieving gigantic size. The huge reptiles died out seventy million years ago, and mammals occupied the earth—they belonged to the Tertiary. According to this scheme, the last mountain uplifts took place at the end of the Tertiary, during the Pliocene; this period lasted until a million years ago, when the Quaternary period, the age of man, began. The Quaternary is also the time of the Ice Age or the Pleistocene—the Paleolithic or Old Stone Age; and the very end of the Quaternary, since the end of the Ice Age, is called Recent time: the Neolithic (Late, or polished, Stone), Bronze, and Iron cultures. Since the appearance of man on earth, or since the beginning of the Ice Age, there have been no uplifts on any substantial scale. In other words, we have been told, the profile of the earth with its mountains and oceans was already established when man first appeared.

In the last few decades, however, numerous facts have emerged from mountains and valleys that tell a different story. In Kashmir, Helmut de Terra discovered sedimentary deposits of an ancient sea bottom that was elevated at places to an altitude of 5000 feet or more and tilted at an angle of as much as 40°; the basin was dragged up by the rise of the mountain. But this was entirely unexpected: "These deposits contain paleolithic fossils." And this, according to Arnold Heim, Swiss geologist, would make it plausible that the mountain passes in the Himalayas may have risen, in the age of man, three thousand feet or more, "however fantastic changes so extensive may seem to a modern geologist."[1]

Studies on the Ice Age in India and Associated Human Cultures, published in 1939 by De Terra, working for the Carnegie Institution, with the assistance of Professor T. T. Paterson of

[1] Arnold Heim and August Gausser, *The Throne of the Gods, an Account of the First Swiss Expedition to the Himalayas* (1939), p. 218.

Harvard University, is one long argument and demonstration that the Himalayas were arising during the Glacial Age and reached their present heights only after the end of the Glacial Age, and actually in historical times. From other mountain ridges came similar reports.

De Terra divided the Ice Age of the Kashmir slopes of the Himalayas into Lower Pleistocene (embracing the first glacial and interglacial stages), Middle Pleistocene (the second, major glaciation and the following interglacial), and Upper Pleistocene (comprising the last two glaciations and an interglacial stage).

"The scenery which this region presented at the beginning of the Pleistocene must have differed greatly from that of our time. . . . The Kashmir valley was less elevated, and its southern rampart, the Pir Panjal, lacked that Alpine grandeur that enchants the traveller today. . . ." Then various formation groups moved "both horizontally and vertically, resulting in a southward displacement of older rocks upon foreland sediments, accompanied by uplift of the mobile belt."[2]

"The main Himalayas suffered sharp uplift in consequence of which the Kashmir lake beds were compressed and dragged upward on the slope of the most mobile range. . . . Uplift was accompanied by a southward shifting of the Pir Panjal block toward the foreland of northwestern India."[3] The Pir Panjal massif that was pushed toward India is at present 15,000 feet high.

In the beginning of this period the fauna was greatly impoverished, but thereafter, judging from remains, large cats, elephants, true horses, pigs, and hippopotami occupied the area.

In the Middle Pleistocene, or Ice Age, there was a "continued uplift." "The archaeological records prove that early paleolithic man inhabited the adjoining plains." De Terra refers to "abundance of paleolithic sites." Man used stone implements of "flake" form, like those found in the Cromer forest-bed in England.

Then once more the Himalayas were pushed upward.

[2] H. de Terra and T. T. Paterson, *Studies on the Ice Age in India and Associated Human Cultures* (1939), p. 223.
[3] Ibid., p. 225.

"Tilting of terraces and lacustrine beds" indicates a "continued uplift of the entire Himalayan tract" during the last phases of the Ice Age.[4]

In the last stages of the Ice Man, when man worked stone in the mountains, he might have been living in the bronze stage down in the valleys. It has been repeatedly admitted by various authorities—quoted subsequently in this book—that the end of the glacial epoch may have been almost contemporaneous with the time of the rise of the great cultures of antiquity, of Egypt and Sumeria and, it follows, also of India, and China. The Stone Age in some regions could have been contemporaneous with the Bronze Age in others. Even now there are numerous tribes in Africa, Australia, and Tierra del Fuego, the southern tip of the Americas, still living in the Stone Age, and many other regions of the modern world would have remained in the Stone Age had it not been for the importation of iron from more advanced regions. The aborigines of Tasmania never got so far as to produce a polished—neolithic—stone implement, and in fact barely entered the crudest stone age. This large island south of Australia was discovered in 1642 by Abel Tasman; the last Tasmanian died in exile in 1876, and the race became extinct.

The more recent uplifts in the Himalayas took place also in the age of modern man. "The postglacial terrace record suggests that there was at least one prominent postglacial advance [of ice]," and this, in the eyes of De Terra and Paterson, is indicative of a diastrophic movement of the mountains. "We must be emphatic on one particular feature—namely, the dependence of Pleistocene glaciation on the diastrophic character of a mobile mountain belt. This relationship, we feel, has not been sufficiently recognized in other glaciated regions, such as Central Asia and the Alps, where similar if not identical conditions are found."[5]

It had been generally assumed that loess—thin windblown dust that is built into clays—is a product of a glacial age. However, in the Himalayas, De Terra reported finding neolithic, or polished stone, implements in loess and commented: "Of importance for us is the fact that loess formation was not restricted to the glacial age but that it continued . . . into postglacial times." In China and in Europe, too, the presence of polished

[4] Ibid., p. 222.   [5] Ibid., p. 223.

stone artifacts in loess prompted a similar revision. The neolithic stage that began, according to the accepted scheme, at the end of the Ice Age, still persisted in Europe and in many other places at the time when, in the centres of civilization, the Bronze Age was already flourishing.

R. Finsterwalder, exploring the Nanga Parbat massif in the western Himalayas (26,660 feet high), dated the Himalayan glaciation as post-glacial; in other words, the expansion of the glaciers in the Himalayas took place much closer to our time than had been previously assumed. Great uplifts of the Himalayas took place in part after the time designated as the Ice Age, or only a few thousand years ago.[6]

Heim, investigating the mountain ranges of western China, adjacent to Tibet and east to the Himalayas, came to the conclusion (1930) that they had been elevated *since* the glacial age.[7]

The great massif of the Himalayas rose to its present height in the age of modern, actually historical, man. "The highest mountains in the world are also the youngest."[8] With their topmost peaks the mountains have shattered the entire scheme of the geology of the "long, long ago."

The Siwalik Hills

The Siwalik Hills are in the foothills of the Himalayas, north of Delhi; they extend for several hundred miles and are 2000 to 3000 feet high. In the nineteenth century their unusually rich fossil beds drew the attention of scientists. Animal bones of species and genera, living and extinct, were found there in most amazing profusion. Some of the animals looked as though nature had conducted an abortive experiment with them and had discarded the species as not fit for life. The carapace of a tortoise twenty feet long was found there; how could such an animal have moved on hilly terrain?[1] The *Elephas ganesa*, an elephant species found in the Siwalik Hills, had tusks about fourteen feet long and over three feet in circumference. One

[6] R. Finsterwalder, "Die Formen der Nanga Parbat-Gruppe," *Zeitschrift der Gesellschaft für Erdkunde zu Berlin*, 1936, pp. 321ff.

[7] Lee, *The Geology of China*, p. 207.

[8] Heim and Gausser, *The Throne of the Gods*, p. 220.

[1] D. N. Wadia, *Geology of India* (2nd ed.; 1939), p. 268.

author says of them: "It is a mystery how these animals ever carried them, owing to their enormous size and leverage."[2]

The Siwalik fossil beds are stocked with animals of so many and such varied species that the animal world of today seems impoverished by comparison. It looks as though all these animals invaded the world at one time: "This sudden bursting on the stage of such a varied population of herbivores, carnivores, rodents and of primates, the highest order of the mammals, must be regarded as a most remarkable instance of rapid evolution of species," writes D. N. Wadia in his *Geology of India*.[3] The hippopotamus, which "generally is a climatically specialized type" (De Terra), pigs, rhinoceroses, apes, oxen filled the interior of the hills almost to bursting. A. R. Wallace, who shares with Darwin the honour of being the originator of the theory of natural selection, was among the first to draw attention, in terms of astonishment, to the Siwalik extinction.

Many of the genera that comprised a wealth of species were extinguished to the last one; some are still represented, but by only a few species. Of nearly thirty species of elephants found in the Siwalik beds, only one species has survived in India. "The sudden and widespread reduction by extinction of the Siwalik mammals is a most startling event for the geologist as well as the biologist. The great carnivores, the varied races of elephants belonging to no less than 25 to 30 species. . . . the numerous tribes of large and highly specialized ungulates [hoofed animals] which found such suitable habitats in the Siwalik jungles of the Pliocene epoch, are to be seen no more in an immediately succeeding age."[4] It used to be assumed that the advent of the Ice Age killed them, but subsequently it has been recognized that great destructions took place in the age of man, much closer to our day.

The older geologists thought that the Siwalik deposits were alluvial in their nature, that they were debris carried down by the torrential Himalayan streams. But it was realized that this explanation "does not appear to be tenable on the ground of the remarkable homogeneity that the deposits possess" and a "uniformity of lithologic composition" in a multitude of isolated

[2] J. T. Wheeler, *The Zonal-Belt Hypothesis* (1908), p. 68. A pair of tusks of this size is on view in the Paleontological musum of Princeton University.
[3] Wadia, *Geology of India*, p. 268.   [4] Ibid., p. 279.

basins, at considerable distances from one another.[5] There must have been some agent that carried these animals and deposited them at the feet of the Himalayas and, after the passage of a geological age, repeated the performance—for in the Siwalik Hills there are animals of more than one age, and signs of more than one destruction. There was also a movement of the ground: "The disrupted part of the fold has slipped bodily over for long distances, thus thrusting the older pre-Siwalik rock of the inner ranges of the mountains over the younger rocks of the outer ranges."[6]

If the cause of these paroxysms and destruction was not local, it must have produced similar effects at the other end of the Himalayas and beyond that range. Thirteen hundred miles from the Siwalik Hills, in central Burma, the deposits cut by the Irrawaddy River "may reach 10,000 feet." "Two fossiliferous horizons occur in this series separated by about 4000 feet of sands." The upper horizon (bed), characterized by mastodon, hippopotamus, and ox, is similar to one of the beds in the Siwaliks. "The sediments are remarkable for the large quantities of fossil-wood associated with them. . . . Hundreds and thousands of entire trunks of silicified trees and huge logs lying in the sandstones" suggest the denudation of "thickly forested" areas.[7] Animals met death and extinction by the elementary forces of nature, which also uprooted forests and from Kashmir to Indo-China threw sand over species and genera in mountains thousands of feet high.

Tiahuanacu in the Andes

In the Andes, at 16° 22' south latitude, a megalithic city was found at an elevation of 12,500 feet, in a region where corn will not ripen. The term "megalithic" fits the dead city only in regard to the great size of the stones in its walls, some of which are flattened and joined with precision. It is situated on the Altiplano, the elevated plain between the Western and Eastern Cordilleras, not far from Lake Titicaca, the largest lake in South America and the highest navigable lake in the world, on the border of Bolivia and Peru.

[5] Ibid., p. 270. [6] Ibid., p. 264. [7] Ibid., pp. 274–75.

"There is a mystery still unsolved on the plateau of Lake Titicaca, which, if stones could speak, would reveal a story of deepest interest. Much of the difficulty in the solution of this mystery is caused by the nature of the region, in the present day, where the enigma still defies explanation." So wrote Sir Clemens Markham in 1910.[1] "Such a region is only capable of sustaining a scanty population of hardy mountaineers and labourers. The mystery consists in the existence of ruins of a great city at the southern side of the lake, the builders being entirely unknown. The city covered a large area, built by highly skilled masons, and with the use of enormous stones."[2]

When the author of the quoted passages posed his question to the scholarly world, Leonard Darwin, then president of the Royal Geographical Society, offered the surmise that the mountain had risen considerably after the city had been built.

"Is such an idea beyond the bonds of possibility?" asked Sir Clemens. Under the assumption that the Andes were once some two or three thousand feet lower than they are now, "maize would then ripen in the basin of Lake Titicaca, and the site of the ruins of Tiahuanacu could support the necessary population. If the megalithic builders were living under these conditions, the problem is solved. If this is geologically impossible, the mystery remains unexplained."[3]

Several years ago another authority, A. Posnansky, wrote in similar vein: "At the present time, the plateau of the Andes is inhospitable and almost sterile. With the present climate, it would not have been suitable in any period as the asylum for great human masses" of the "most important prehistoric centre of the world."[4] "Endless agricultural terraces" of the people who lived in this region in pre-Inca days can still be recognized. "Today this region is at a very great height above sea level. In remote periods it was lower."[5]

The terraces rise to a height of 15,000 feet, twenty-five hundred feet above Tiahuanacu, and still higher, up to 18,400 feet above sea level, or to the present line of eternal snow on Illimani. The conservative view among evolutionists and geologists is

[1] Clemens Markham, *The Incas of Peru* (1910), p. 21.
[2] Ibid., p. 23.   [3] Ibid.
[4] A. Posnansky, *Tiahuanacu, the Cradle of the American Man* (1945), p. 15.
[5] Ibid., pp. 1, 39.

that mountain making is a slow process, observable in minute changes, and that because it is a continuous process there never could have been spontaneous upliftings on a large scale. In the case of Tiahuanacu, however, the change in altitude apparently occurred after the city was built, and this could not have been the result of a slow process that required hundreds of thousands of years to produce a visible alteration.

Once Tiahuanacu was at the water's edge; then Lake Titacaca was ninety feet higher, as its old strand line discloses. But this strand line is tilted and in other places it is more than 360 feet above the present level of the lake. There are numerous raised beaches; and stress was put on "the freshness of many of the strandlines and the modern character of such fossils as occur."[6]

Further investigation into the topography of the Andes and the fauna of Lake Titacaca, togther with a chemical analysis of this lake and others on the same plateau, established that the plateau was at one time at sea level, or 12,500 feet lower than it is today. "Titicaca and Poopo, lake and salt bed of Coipaga, salt beds of Uyuni—several of these lakes and salt beds have chemical compositions similar to those of the ocean."[7] As long ago as 1875 Alexander Agassiz demonstrated the existence of a marine crustaceous fauna in Lake Titicaca.[8] At a higher elevation the sediment of an enormous dried-up lake, whose waters were almost potable, "is full of characteristic molluscs, such as Paludestrina and Ancylus, which shows that it is, geologically speaking, of relatively modern origin."[9]

Sometime in the remote past the entire Altiplano with its lakes rose from the bottom of the ocean. At some other time point a city was built there and terraces were laid out on the elevation around it; then in another disturbance the mountains were thrust up and the area became uninhabitable.

The barrier of the Cordilleras that separates the Altiplano from the valley to the east was torn apart and gigantic blocks were thrown into the chasm. Lyell, combating the idea of a

[6] H. P. Moon, "The Geology and Physiography of the Altiplano of Peru and Bolivia," *The Transactions of the Linnean Society of London*, 3rd Series, Vol. I, Pt. 1 (1939), p. 32.
[7] Posnansky, *Tiahuanacu*, p. 23.
[8] *Proceedings of the American Academy of Arts and Sciences*, 1876.
[9] Posnansky, *Tiahuanacu*, p. 23.

universal flood, offered the theory that the bursting of the Sierra barrier opened the way for a large lake on the Altiplano, which cascaded down into the valley and caused the aborigines to create the myth of a universal flood.[10]

Not so long ago an explanation of the mystery of Lake Titicaca and of the fortress Tiahuanacu on its shore was put forward in the light of Hörbiger's theory: A moon circled very close to the earth, pulling the waters of the oceans toward the equator; by its gravitational pull, the moon held, day and night, the water of the ocean at the altitude of Tiahuanacu: "The level of the ocean must have been at least 13,000 feet higher."[11] Then the moon crashed into the earth, and the oceans receded to the poles, leaving the island with its megalithic city as a mountain above the sea bottom, now the continent of the tropical and subtropical Americas. All this happened millions of years before our moon was caught by the earth, and thus the ruins of the megalithic city Tiahuanacu are millions of years old, that is, the city must have been built long "before the Flood."

This theory is bizarre. The geological record indicates a late elevation of the Andes, and the time of its origin is brought ever closer to our time. Archaeological and radio-carbon analyses indicate that the age of the Andean culture and of the city is not much older than four thousand years.[12] Not only the "built before the Flood" theory collapses; so does the belief that the last elevation of the Andes was in the Tertiary, or more than a million years ago.

Sometime in the remote past the Altiplano was at or below sea level, so that originally its lakes were part of a sea gulf. The last upheaval, however, took place in an early historical period, after the city of Tiahuanacu had been built; the lakes were dragged up, and the Altiplano and the entire chain of the Andes rose to their present height.

The ancient stronghold of Ollantaytambo in Peru is built on top of an elevation; it is constructed of blocks of stone twelve to eighteen feet high. "These Cyclopean stones were hewn from

[10] Lyell, *Principles of Geology*, I, 89; III, 270.
[11] H. S. Bellamy, *Built before the Flood: The Problem of the Tiahuanacu Ruins* (1947), p. 14.
[12] F. C. Hibben, *Treasure in the Dust* (1951), p. 56.

the quarry seven miles away. . . . How the stones were carried down to the river in the valley, shipped on rafts, and carried up to the site of the fortress remains a mystery archaeologists cannot solve."[13]

Another fortress or monastery, Ollantayparubo, in the Urubamba Valley in Peru, northwest of Lake Titicaca, "perches upon a tiny plateau some 13,000 feet above sea-level, in an uninhabitable region of precipices, chasms, and gorges." It is built of red porphyry blocks. The blocks must have been brought "from a considerable distance . . . down steep slopes, across swift and turbulent rivers, and up precipitous rock-faces which hardly allow a foothold."[14] It has been suggested that the transportation of the building blocks was feasible only if the topography of these localities was different at the time of the construction. However, definite proof in this connection is lacking, and changes in topography must be deduced from abandoned terraces, from molluscs of the dried-up lakes, from tilted shorelines, and from other similar indications.

Charles Darwin, on his travels in South America in 1834–35, was impressed by the raised beaches at Valparaiso, Chile, at the foot of the Andes. He found that the former surf line was at an altitude of 1300 feet. He was impressed even more by the fact that the sea shells found at this altitude were still undecayed, to him a clear indication that the land had risen 1300 feet from the Pacific Ocean in a very recent period, "within the period during which upraised shells remained undecayed on the surface."[15] And since only a few intermediary surf lines can be detected, the elevation could not have proceeded little by little.

Darwin also observed that "the excessively disturbed condition of the strata in the Cordillera, so far from indicating single periods of extreme violence, presents insuperable difficulties, except on the admission that the massses of once liquefied rocks of the axes were repeatedly injected with intervals sufficiently long for their successive cooling and consolidation."[16]

At present it is the common view that the Andes were created, not so much by compression of the strata, as by magma,

[13] Don Ternel, in *Travel*, April 1945.
[14] Bellamy, *Built before the Flood*, p. 63.
[15] Charles Darwin, *Geological Observations on the Volcanic Islands and Parts of South America*, Pt. II, Chap. 15.
[16] Ibid.

or molten rock, invading the strata and lifting them. The Andes also abound in volcanoes, some exceedingly high and enormously large.

The foothills of the Andes hide numerous deserted towns and abandoned terraces, monuments to a vanished civilization. The terraces that go up the slopes of the Andes and reach the eternal snow line and continue under the snow to some unidentified altitude prove that it was not a conqueror nor a plague that put the seal of death on gardens and towns. In Peru "aerial surveys in the dry belt west of the Andes have shown an unexpected number of old ruins, and an almost incredible number of terraces for cultivation."[17]

When Darwin mounted the Uspallata Range, 7000 feet high in the Andes, and looked down on the plain of Argentina from a little forest of petrified trees broken off a few feet above the ground, he wrote in his *Journal*:

"It required little geological practice to interpret the marvellous story which this scene at once unfolded; though I confess I was at first so much astonished that I could scarcely believe the plainest evidence. I saw the spot where a cluster of fine trees once waved their branches on the shores of the Atlantic, when that ocean—now driven back 700 miles—came to the foot of the Andes. I saw that they had sprung from a volcanic soil which had been raised above the level of the sea, and that subsequently this dry land, with its upright trees, had been let down into the depths of the ocean. In these depths, the formerly dry land was covered by sedimentary beds, and these again by enormous streams of submarine lava—one such mass attaining the thickness of a thousand feet; and these deluges of molten stone and aqueous deposits five times alternately had been spread out. The ocean which received such thick masses, must have been profoundly deep; but again subterranean forces exerted themselves, and now I beheld the bed of that ocean, forming a chain of mountains more than seven thousand feet in height. . . . Vast, and scarcely comprehensible as such changes must ever appear, yet they have all occurred within a period, recent when

[17] E. Huntington, "Climatic Pulsations" in *Hylluingsskrift*, dedicated to Sven Hedin (1935), p. 578.

compared with the history of the Cordillera; and the Cordillera itself is absolutely modern as compared with many of the fossiliferous strata of Europe and America."[18]

But how extremely young the Cordillera of the Andes is, only the research of recent years has brought out.

The Columbia Plateau

Great quantities of lava "flowed out in Washington, Oregon and Idaho, where some two hundred thousand square miles were covered to depths of hundreds and even several thousands of feet. The Snake River has cut the Seven Devils canyon more than three thousand feet deep without reaching the bottom of the lavas."[1]

This enormous area, embracing all the Northern states between the Rocky Mountains and the Pacific coast, was flooded with molten rock and metal pouring out of fissures torn in the ground. Certainly it does not look like a volcanic eruption of our days, and for this reason alone, if not for a multitude of others, the principle of uniformity is definitely misleading.

The depth of the lava of this vast Columbia Plateau is "as great as 5000 feet or more."[2] Even on the supposition that it was ejected in paroxysms, each time spreading a sheet only seventy-five feet thick, it is still enormous, and then such an ejection must have been repeated as much as seventy times in the Cenozoic Age—the age of mammals and man.

And here is a striking thing, striking because we are too readily disposed to consider that we have solved a problem when we remove it to the remote past. "All competent observers have remarked the freshness of lava deposits in the Snake River valley in Idaho."[3]

Only a few thousand years ago lava flowed there over an area larger than France, Switzerland, and Belgium combined; it flowed not as a creek, not as a river, not even as an overflowing

[18] *Journal of Researches . . . during the Voyage of H.M.S. Beagle.* From the entry of March 30, 1835.

[1] Chamberlin, in *The World and Man,* ed. Moulton, p. 85.

[2] W. J. Miller, *An Introduction to Historical Geology* (5th ed., 2nd printing; 1946), p. 355.

[3] Wright, *The Ice Age in North America,* p. 688.

stream, but as a flood, deluging horizon after horizon, filling all the valleys, devouring all the forests and habitations, steaming large lakes out of existence as though they were little potholes filled with water, swelling ever higher and overtopping mountains and burying them deep beneath molten stone, boiling and bubbling, thousands of feet thick, billions of tons heavy.

In 1889, on the occasion of the boring of an artesian well at Nampa, Idaho, on the Columbia Plateau near the Snake River, a small figurine of baked clay was extracted from a depth of 320 feet, penetrated after piercing a sheet of basalt lava fifteen feet thick. G. F. Wright described the find and wrote: "The well was tubed with heavy iron tubing six inches in diameter, so that there could be no mistake about the occurrence of the image at the depth stated." He also added: "No one has come forward to challenge the evidence except on purely a priori grounds arising from preconceived opinions of the extreme antiquity of the deposits."[4]

Before the last lava sheets spread over the Columbia Plateau there were human abodes in the area.

A Continent Torn Apart

"Africa was in tension and torn by north and south fractures [which] along with the sinking of a strip of the crust formed the longest meridional land valley on earth. . . . From Lebanon [in Syria], then, almost to the Cape there runs a deep and comparatively narrow valley, margined by almost vertical sides, and occupied by the sea, by salt steppes and old lake basins, and by a series of over twenty lakes, of which only one has an outlet to the sea. This is a condition of things absolutely unlike anything else on the surface of the earth."[1] The author of these lines, J. W. Gregory, the famous explorer of the Great African Rift, adopted the view that a general common cause created the entire Rift from its north to its south end.

The Rift begins in the valley of the Orontes River in Syria;

[4] Ibid., pp. 701–3.
[1] J. W. Gregory, "Contributions to the Physical Geography of British East Africa," *Geographical Journal*, IV (1894), 290.

at Baalbek it goes over to the Litani River Valley, then to Lake Huleh in Palestine; along the Jordan River to the Sea of Galilee, called also Gennesaret or the Sea of Tiberias, which lies in a depression below the level of the Mediterranean; from there to the Dead Sea, the deepest depression on earth, between the Judean and Moabite mountainous plateaux that were torn apart; then along the Araba Valley to the Gulf of Aqaba in the Red Sea and across the channel of this sea into Africa; thence for an enormous distance to the Sabie River in the Transvaal, branching, on the way, eastward to the Gulf of Aden and westward to Tanganyika and the Upper Nile, and the rift valleys of Lakes Moeris and Upemba in the central Congo—all the way from about 36° north latitude in Syria to about 28° south latitude in East Africa, in a sinuous line along a meridian for more than a third of the way from one pole to another.

It was recognized that a horizontal force of one kind or another had been the cause of this rift valley. "The simplest and earliest thought was that Africa had been pulled apart."[2] However, another school of geologists questioned whether the Rift could not have been produced under horizontal pressure, which forced the margins of the rift valley up and the valley strip down. After a long debate the consensus restated the view expressed by Eduard Suess, a prominent geologist at the turn of the century: "The opening of fissures of such magnitude can be explained only by the action of a tension, directed perpendicularly to the trend of the split, the tension being relieved in the instant of bursting, that is, of opening of the fissure."[3] He observed also that immense floods of lava gushed out of the earth along the Rift and a most vigorous volcanic action took place. Suess brought to geology the now generally accepted concept of Gondwana land, a continental mass that occupied the larger portion of the Indian Ocean, and that in a relatively recent subsidence was torn apart and drowned. The subsidence of the Gondwana continent could have caused a strain on western

[2] B. Willis, *East African Plateaus and Rift Valleys* (1936), p. 1.
[3] Ibid., p. 13. E. Krenkel, a German authority, wrote in *Die Bruchzonen Ostafrikas* (1922): "The tectonic setting of the East African fault zones, whether considered in detail or as a whole, admits of only one explanation: they are zones of tearing apart of the crust, produced by a direct tension. . . . The action of compressive forces is nowhere recognizable." (Trans. B. Willis.)

Asia and Africa, and under this tension the land must have been rent and the Great Rift formed.

Gregory wrote: "The nearest approach in size [to the Rift] can probably be found on the moon, whose clefts or rills no doubt represent long, steeply walled valleys and present to us much the same aspects as this East African valley would do to any inhabitants of our satellite. Not the least interesting of the points raised by the African-Red-Sea-Jordan depression is the possibility that it may explain the nature of those lunar clefts which have so long been a puzzle to astronomers."[4]

The Rift was produced by tension; hence the rifts on the moon were also caused by tension. Gregory followed Suess in linking the Great Rift Valley with "the mountain chains due to the last great uplift of fold mountains" in Europe, Asia, and the Americas. Thus the time of the last uplift, if established, would also clarify the time when Africa suffered the Great Rift. It is probable, too, that the Rift began in one great tension and increased in the next.

Gregory concluded: "This wide-spread valley system is obviously not the result of some local fracture. Its length is about one-sixth of the circumference of the Earth. It must have some world-wide cause, the first promising clue to which is the date of its formation."[5]

Although Gregory thought that the Rift first came into being at an early epoch—because of marine fossils found in it—he also saw signs of great earth movements along the Rift "at a recent date." "Some of the fault-scarps are so bare and sharp that they must be of very recent date. This continuation of earth-movements into the human period is one of the most striking features of the district." Gregory found also that human memory retained recollection of the upheaval. "All along the line the natives have traditions of great changes in the structure of the country."[6]

The globe was in tension and its crust cracked along a meridian for most of the length of the African continent. The cause may have been the subsidence of the Indian Ocean, or both

[4] Gregory, *Geographical Journal*, IV (1894); *The Great Rift Valley* (1896), p. 6.
[5] Gregory, "The African Rift Valleys," *Geographical Journal*, LVI (1920), 31ff.
[6] Gregory, *The Great Rift Valley*, pp. 5, 236.

tension in Africa and subsidence in the Indian Ocean could have a common cause. The mountain ridge on the floor of the Atlantic Ocean may have been produced by the same cause; and the time of the rupture and faulting must have been coincident with one of the periods of mountain formation in Europe and Asia. Those mountains attained their present height in the age of man; the Rift, it is assumed today, was also created largely in the age of man at the end of the Ice Age.[7]

What kind of force is necessary to tear apart a continent? Whence came the tension that was relieved by the bursting of the African land mass? Ice did not do it, nor the wind that erodes mountain heights, nor the rivulets that carry eroded detritus down to the sea.

[7] Flint, *Glacial Geology*, p. 523: "Late-Pleistocene mountain uplift occurred in the Himalayan region and in the Alps, and large-scale rifting took place in eastern Africa."

Chapter VII

DESERTS AND OCEANS

The Sahara

The Sahara Desert, which stretches from the Nile to the Atlantic Ocean across the continent of Africa and covers 3,500,000 square miles, about the area of all of Europe, is the greatest desert on earth. What is now the desert of Sahara was an open grassland or steppe in earlier days. Drawings on rock of herds of cattle, made by early dwellers in this region, were discovered by Barth in 1850. Since then many more drawings have been found. The animals depicted no longer inhabit these regions, and many are generally extinct. It is asserted that the Sahara once had a large human population that lived in vast green forests and on fat pasture lands. Neolithic implements, vessels and weapons made of polished stone, were found close to the drawings. Such drawings and implements were discovered in the eastern as well as the western Sahara. Men lived in these "densely populated" (Flint) regions and cattle pastured where today enormous expanses of sand stretch for thousands of miles.

Several theories have been offered to explain the prodigious quantity of sand in the Sahara. "The theory of marine origin is now no longer tenable."[1] The sand, it was found, is of recent origin. It is assumed that when a large part of Europe was under ice the Sahara was in a warm and moist temperate zone; later the soil lost its moisture and the rock crumbled to sand when left to the mercy of the sun and the wind.

How long ago was it that conditions in the Sahara were suitable for human occupation? Movers, the noted Orientalist of

[1] "Sahara," *Encyclopaedia Britannica* (14th ed.), Vol XIX.

the last century, author of a large work on the Phoenicians, decided that the drawings in the Sahara were the work of the Phoenicians.[2] It was likewise observed that on the drawings discovered by Barth the cattle wore discs between their horns, just as in Egyptian drawings.[3] Also, the Egyptian god Set was found pictured on the rocks. And there are rock paintings of war chariots drawn by horses "in an area where these animals could not survive two days without extraordinary precautions."[4]

The extinct animals in the drawings suggest that these pictures were made sometime during the Ice Age; but the Egyptian motifs in the very same drawings suggest that they were made in historical times.

The conflict between the historical and the paleontological evidence, and of both of them with the geological evidence, is resolved if one or more catastrophes intervened. It appears that a large part of the region was occupied by an inland lake, or vast marsh, known to the ancients as Lake Triton. In a stupendous catastrophe the lake emptied itself into the Atlantic, and the sand on its bottom and shores was left behind, forming a desert when tectonic movements sealed off the springs that fed the lake. The "land of pastures and forests" became a desert of sand; hippopotami that live in water and elephants disappeared and with them also the hunter and the farmer.

The French savant A. Berthelot says: "It is possible that Stone Age man witnessed in Africa three notable events: the sinking of the Spanish-Atlas chain that opened the Strait of Gibraltar and created a junction between the Mediterranean Sea and the Ocean; the collapse that cut off the Canary Islands from the African continent; the opening of the Strait of Bab-el-Mandeb, separating Arabia from Ethiopia."[5] Berthelot, however, ascribed these great tectonic changes to the time of prehistoric man and Abbé Breuil actually showed that prehistoric man already occupied these regions as the eolithic or very crudely chipped stone artifacts indicate. But at a later date people of advanced culture, contemporary with pharaonic Egypt, lived in communities, pastured their cattle, and left

[2] L. Frobenius and Douglas C. Fox, *Prehistoric Rock Pictures in Europe and Africa* (Museum of Modern Art, 1937), p. 38.
[3] Ibid., pp. 39–40. [4] P. LeCler, *Sahara* (1954), p. 46.
[5] A. Berthelot, *L'Afrique saharienne et soudanaise* (1927), p. 85.

their tools and drawings there. Then in an upheaval, of which many traditions persist in classical literature, the Atlas Mountains were torn apart, the great lake was emptied, and the watery region became the great and awesome desert—the Sahara.

Arabia

There is a "certainty beyond challenge that when the ice-cap of the last Glacial period covered a large part of the northern hemisphere, at least three great rivers flowed from west to east across the whole width of the [Arabian] Peninsula." So wrote Philby in his book *Arabia*.[1] There was also a large lake in Arabia that disappeared in some geological or climatal change.[2]

At present, from Palmyra to Mecca and beyond, the Arabian Peninsula is a waterless desert, interspersed with volcanoes active not so long ago, but now extinct, the last eruption having taken place in 1253.[3] There were also, sometime in the past, numerous geysers all likewise extinct now.

Twenty-eight fields of burned and broken stones, called harras, are found in Arabia, mostly in the western half of the great desert. Some single fields are one hundred miles in diameter and occupy an area of six or seven thousand square miles, stone lying close to stone, so densely packed that passage through the field is almost impossible.[4] The stones are sharp-edged and scorched black. No volcanic eruption could have cast scorched stones over fields as large as the harras; neither would the stones from volcanoes have been so evenly spread. The absence, in most cases, of lava—the stones lie free—also speaks against a volcanic origin of the stones.

It appears that the blackened and broken stones of the harras are trains of meteorites, scorched in their passage through the atmosphere, that broke during their fall, as bolides do, or on

[1] H. St. J. B. Philby, *Arabia* (1930), p. xv.
[2] Described by Bertram Thomas; cf. C. P. Grant, *The Syrian Desert* (1937), p. 53.
[3] B. Moritz, *Arabien, Studien zur physikalischen und historischen Geographie des Landes* (1923).
[4] Described by C. M. Doughty and by B. Moritz. The latter's book, *Arabien*, contains a close-up photograph of a harra.

reaching the ground. Billions of stones in a single harra indicate that the trains of meteorites were very large and can be classed as comets. Despite alternate exposure to the thermal action of the hot desert sun and the cool desert night, the sharp edges of the stones have been preserved, which shows that they fell in a not too distant period of time. Following the procedure adopted in this book, literary references to the harras of Arabia in ancient Hebrew and Arabic literatures will not be dealt with here.

Meteorites that fall on the earth are of two kinds. One consists of iron with an admixture of nickel; by means of this admixture and the characteristic pattern seen in the cut surface of such stones, their meteoric origin can be easily established. The other group, probably larger than the first, does not differ in its composition from the rocks of the earth and cannot be distinguished unless the fall has been observed, or, as in the case of the stones of the harras, their scorched and broken condition, together with their occurrence in large fields, speak for their extraterrestrial origin.

Larger bodies than the stones of the harras fell on Arabia, too. In Wobar in the desert there is a meteoric crater with meteoric iron and silica glass spread around it.[5]

Large rivers that disappeared, numerous volcanoes that burned and were extinguished, blackened stones that fell in areas each of them a hundred times larger than any volcanic eruption could have covered, and meteoric iron spread around a large crater—all of these bespeak great upheavals in nature in recent as well as earlier ages, to which the vast peninsula of Arabia was more than once subjected.

In the southern part of the great Arabian desert, ancient ruins, almost entirely obliterated by time and the elements, and vestiges of cultivation are silent witnesses of the time when the land there was hospitable and fruitful; it was as copiously watered and luxuriously forested as India on the same latitude. Orchards covered Hadhramaut and Aden. It was a land of plenty, paradise on earth, but following a sudden catastrophe,

[5] R. Schwinner, *Physikalische Geologie* (1936), I, 114, 163; L. J. Spencer, "Meteoric Iron and Silica Glass from the Craters of Henbury (Central Australia) and Wobar (Arabia)," *Mineralogical Magazine*, XXIII (1933), 387–404.

Arabia Felix turned to a barren land. Arabia Petraea, the western part of the desert, is a dusty rock of lava that is broken by the Great Rift, with the Dead Sea, an inner lake, on its bottom. Sulphurous springs flow into it, and asphalt rises from its floor and floats on it.

Like the Sahara and Arabian deserts, other great deserts of the world disclose the fact that they were inhabited and cultivated sometime in the past. On the Tibetan plateau and in the Gobi Desert remains of early prosperous civilizations were found with occasional ruins surviving from those times when the great barren tracts were cultivated. In the Gobi Desert, as in the Arabian and Sahara deserts, the impression is gained that in a tectonic disturbance the subterranean water dropped to a great depth, the sources became sealed, and the rivers dried up completely. Some changes in ground structure or in ground currents also affect the clouds, which pass over such lands without unburdening themselves.

The Carolina Bays

Peculiar elliptical depressions, or "oval craters," locally called "bays," are thickly scattered over the Carolina coast of the United States and more sparsely over the entire Atlantic coastal plain from southern New Jersey to northeastern Florida. These marshy depressions are numbered in the tens of thousands and, occording to the latest estimate, their number may reach half a million.[1] Measurements made on more prominent ones, seaward from Darlington, show that the larger bays average 2200 feet in length, and in single cases exceed 8000 feet. A remarkable feature of these depressions is their parallelism: the long axis of each of them extends from northwest to southeast, and the precision of the parallelism is "striking." Around the bays are rims of earth, invariably elevated at the southeastern end. These oval depressions may be seen especially well in aerial photographs. Any theory as to their origin must explain their form, the ellipticity of which increases with the size of the

---

[1] Douglas Johnson, *The Origin of the Carolina Bays* (1942); W. F. Prouty, "Carolina Bays and Their Origin," *Bulletin of the Geological Society of America*, LXIII (1952), 167–224.

bays; their parallel alignment; and the elevated rims at their southeastern ends.

In 1933 a theory was presented by Melton and Schriever of the University of Oklahoma, according to which the bays are scars left by a "meteoric shower or colliding comet."[2] Since then the majority of the authors who have dealt with the problem have accepted this view, and it has found its way into textbooks as the usual interpretation.[3] The authors of the theory stress the fact that, "Since the origin of the bays apparently cannot be explained by the well-known types of geological activity, an extraordinary process must be found. Such a process is suggested by the elliptical shape, the parallel alignment, and the systematic arrangement of elevated rims."

The comet must have struck from the northwest. "If the cosmic masses approached this region from the north-west the major axes would have the desired alignment." The time when the catastrophe took place was estimated as sometime during the Ice Age. The bays are "filled to a considerable extent by the deposition of sand and silt, a process which doubtless occurred while the region was covered by the sea during the terrace-forming marine invasion of the Pleistocene [glacial] period."[4] But the possibility was also envisaged that "the collision took place" through "the shallow ocean water during the marine invasion." The swarm of meteorites must have been large enough to hit an area from Florida to New Jersey.

Some critics disagree with the idea that the bays originated in the Ice Age or "are relatively ancient," and place their origin in a more recent time.[5] The craters were produced by meteoric impact, either by direct hits or by explosion in the air close to the ground, thus cuasing the formation of vast numbers of depressions. Some of the bays, it is assumed, are on the bottom of the ocean. It was also stressed that "a very large number of meteorites have been discovered in the southern Appalachian region, in Virginia, North and South Carolina, Georgia, Alabama, Kentucky, and Tennessee."[6]

[2] F. A. Melton and W. Schriever, "The Carolina Bays—Are They Meteorite Scars?" *Journal of Geology*, XLI (1933).
[3] Cf. Johnson, *The Origin of the Carolina Bays*, p. 4.
[4] Melton and Schriever, *Journal of Geology*, XLI (1933), 56.
[5] Johnson, *The Origin of the Carolina Bays*, p. 93.
[6] Cf. C. P. Olivier, *Meteors* (1925), p. 240.

The Bottom of the Atlantic

In the autumn of 1949, Professor M. Ewing of Columbia University published a report on an expedition to the Atlantic Ocean. Explorations were carried on especially in the region about the Mid-Atlantic Ridge, the mountainous chain that runs from north to south, following the general outlines of the ocean. The Ridge, as well as the ocean bottom to the west and to the east, disclosed to the expedition a series of facts that amount to "new scientific puzzles."[1]

"One was the discovery of prehistoric beach sand . . . brought up in one case from a depth of two and the other nearly three and one half miles, far from any place where beaches exist today." One of these sand deposits was found twelve hundred miles from land.

Sand is produced from rocks by the eroding action of sea waves pounding the coast, and by the action of rain and wind and the alternation of heat and cold. On the bottom of the ocean the temperature is constant; there are no currents; it is a region of motionless stillness. Mid-ocean bottoms are covered with ooze made up of silt so fine that its particles can be carried suspended in ocean water for a long time before they sink to the bottom, there to build sediment. The ooze contains skeletons of the minute animals, foraminifera, that live in the upper waters of the ocean in vast numbers. But there should be no coarse sand on the mid-ocean floor, because sand is native to land areas and to the continental shelf, the coastal rim of the ocean and its seas.

These considerations presented Professor Ewing with a dilemma: "Either the land must have sunk two to three miles, or the sea once must have been two to three miles lower than now. Either conclusion is startling. If the sea was once two miles lower, where could all the extra water have gone?"

It is regarded as an accepted truth in geology that the seas have not changed their beds with the exception of encroachment by shallow water on depressed continental areas. Thus it was difficult to accept the startling conclusion that the bottom of the ocean was at some time in the past dry land.

[1] M. Ewing, "New Discoveries on the Mid-Atlantic Ridge," *National Geographic Magazine*, Vol. XCVI, No. 5 (November 1949).

But there was another surprise in store for the expedition. The thickness of the sediment on the ocean bottom was measured by the well-developed method of sound echoes. An explosion is set off and the time it takes for the echo to return from the sediment on the floor of the ocean is compared with the time required for a second echo to return from the bottom of the sediment, or from the bedrock, basalt or granite. "These measurements clearly indicate thousands of feet of sediments on the foothills of the Ridge. Surprisingly, however, we have found that in the great flat basins on either side of the Ridge, this sediment appears to be less than 100 feet thick, a fact so startling . . ." Actually, the echoes arrived almost simultaneously, and the most that could be attributed in such circumstances to the sediment was less than one hundred feet of thickness, or the margin of error.

"Always it had been thought the sediment must be extremely thick, since it had been accumulating for countless ages. . . . But on the level basins that flank the Mid-Atlantic Ridge our signals reflected from the bottom mud and from bedrock came back too close together to measure the time between them. . . . They show the sediment in the basins is less than 100 feet thick."

The absence of thick sediment on the level floor presents "another of many scientific riddles our expedition propounded." It indicates that the bottom of the Atlantic Ocean on both sides of the Ridge was only very recently formed. At the same time, on the flanks of the Ridge the layers of sediment in some places are "thousands of feet thick, as was expected."

"These ocean-bottom sediments we measured are formed from the shells and skeletons of countless small sea creatures" and "from volcanic dust and wind-blown soil drifting out over the sea; and from the ashes of burned out meteorites and cosmic dust from outer space sifting constantly down upon the earth."

Burned-out meteorites and cosmic dust elicited the question: If the meteoric dust in our age is so sparse that it is hardly detectable on the snow of high mountains, how could ashes of burned-out meteorites and cosmic dust make up a substantial part of the oceanic sediment? And how could it be that all other sources, including detritus carried by rivers, have created in all ages since the beginning a sediment of only very moderate thickness?

"We dredged up rocks of igneous, or 'fire-made,' type from the sides and tops of peaks on the Mid-Atlantic Ridge, which indicated that submarine volcanoes and lava flows have been active there. Probably the whole Ridge is highly volcanic, with perhaps thousands of lava outpourings and active and extinct cones scattered along its entire length."

And not only the submarine Ridge is volcanic. "There are many peaks of volcanic origin scattered over the Atlantic Basin." In the direction of the Azores the expedition found an uncharted submarine mountain, 8000 feet high, with "many layers of volcanic ash," and farther on, a great hole dropping down 1809 fathoms (10,854 feet), "as if a volcano had caved in there at some time in the past."

Lava flowed under the water of the ocean, and the water must have boiled; meteorites, ashes, and cosmic dust fell from the sky; land was submerged thousands of fathoms deep, and beaches sank over three miles into the depths.

From the abyss of the ocean, rocks marked with deep scratches were raised by the expedition. "In a depth of 3600 feet (600 fathoms) we found rocks that tell an interesting story about the past history of the Atlantic Ocean . . . granite and sedimentary rocks of types which originally must have been part of a continent. Most of the rocks that we dredge here were rounded and marked with deep scratches, or striations." Such marks on rocks are regularly ascribed to the action of glaciers that held rocks in a firm grip and moved them over the surface of other rocks. "But we also found some loosely consolidated mud stones, so soft and weak they would not have held together in the iron grasp of a glacier. How they got out here is another riddle to be solved by further research."

Finally, the very entrance to New York Harbour, the Hudson River, was found to have a canyon running into the ocean, not only for the width of the continental shelf, a hundred and twenty miles offshore, as has been known for some time, but also for another hundred miles in deeper water. "If all this valley was originally carved out by the river on dry land, as seems probable, it means either that the ocean floor of the Eastern seaboard of North America once must have stood about two miles above its present level and has since subsided, or else that the level of the sea was once about two miles lower than now."[2] Each one of these possibilities indicates an upheaval.

All in all, the results of the expedition of the summer of 1949 strongly indicate that, at some time not so long ago, in numerous places where the Atlantic Ocean is today there were land and beaches, and that in revolutions on a great scale land became sea thousands of fathoms deep. The leader of the *Atlantis* expedition, whom we have quoted here, did not use the term "revolution," but it is unavoidable in the face of the expedition's finds. In order not to be regarded as the proponent of a heresy, Ewing made only a negative statement: "There is no reason to believe that this mighty underwater mass of mountains is connected in any way with the legendary lost Altantis which Plato described as having sunk beneath the waves."

The Floor of the Seas

In July 1947 a Swedish deep-sea expedition left Göteborg on the *Albatross* for a fifteen-month journey around the world to investigate the bottom of the seas on the seventeen thousand miles of the ship's course with the help of a newly constructed vacuum core sampler. In the sediment that covers the rocky bottom of the oceans the expedition found, in the words of its leader, H. Pettersson, director of the Oceanographic Institute at Göteborg, "evidence of great catastrophes that have altered the face of the earth."[1]

"Climatic catastrophes, which piled thousands of feet of ice on the higher latitudes of the continents, also covered the oceans with icebergs and ice fields at lower latitudes and chilled the surface waters even down to the Equator. Volcanic catastrophes cast rains of ash over the sea." This ash is preserved in the sedimentary bottom of the oceans. "Tectonic catastrophes raised or lowered the ocean bottom hundreds and even thousands of feet,

[2] Ibid.
[1] Pettersson, in advance of the detailed report of the expedition, gave a popular account in an article entitled "Exploring the Ocean Floor," *Scientific American*, August 1950.

spreading huge 'tidal' waves which destroyed plant and animal life on the coastal plains."

At many places, such as the coast of Sweden, the bottom of the sea proved to consist of "a lava bed of geologically recent origin, covered only by a thin veneer of sediment. . . . The sediments of the Pacific and Indian Oceans, which often bore particles of volcanic material, also testified to the importance of vulcanism in submarine geology. Some of our cores from the Mediterranean were marked with corase-grained layers consisting largely of volcanic ash that had settled on the bottom after great volcanic explosions. These layers are an unrivalled record of the irregular volcanic activity of the past."

The oceanic floor all around the globe bears witness that the oceans of the earth were the scenes of repeated violent catastrophes when flows of lava and volcanic ash covered the precipitously rising or falling bedrock and tidal waves raced against continents.

The bottom of the seas and oceans also contains evidence that the earth was showered with meteorites on a very large scale. In many places the bottom consists of red clay. Samples of the red clay from the central Pacific showed a "surprisingly high content of nickel," and also a high content of radium, though the water of the ocean is almost completely free of these elements.[2] The red clay is red because it contains ferruginous (iron) compounds. Meteoric iron differs from iron of terrestrial origin in its admixture of nickel, and it is this characteristic that makes it possible to differentiate iron tools of early ages, for instance of the pyramid age in Egypt, and to decide whether iron pieces were smelted from ore or were worked meteorites. "Nickel is a very rare element in most terrestrial rocks and continental sediments, and it is almost absent from the ocean waters. On the other hand, it is one of the main components of meteorites."[3]

Thus it is assumed that the origin of the abysmal nickel was in meteoric dust or "the very heavy showers of meteors in the remote past. The principal difficulty of this explanation is that it requires a rate of accretion of meteoric dust several hundred times greater than that which astronomers, who base their

[2] Pettersson, "Chronology of the Deep Ocean Bed," *Tellus* (*Quarterly Journal of Geophysics*), I, 1949.

[3] Pettersson, *Westward Ho with the Albatross* (1953), pp. 149–50.

estimates on visual and telescopic counts of meteors, are presently prepared to admit."[4]

In a later publication, a popularized account of the *Albatross* expedition, Pettersson writes: "Assuming the average nickel content of meteoric dust to be two per cent, an approximate value for the rate of accretion of cosmic dust to the whole Earth can be worked out from these data. The result is very high—about 10,000 tons per day, or over a thousand times higher than the value computed from counting the shooting stars and estimating their mass."[5]

In other words at some time or times there was such a fall of meteoric dust that, apportioned throughout the entire age of the ocean, it would increase a thousandfold the daily accumulation of meteoric dust since the birth of the ocean.

The ash and lava on the bottom of the oceans indicate catastrophic occurrences in the past. Iron and nickel point to celestial showers of meteorites, and thus possibly also to the cause of the tectonic ruptures, of the collapse of the ocean floor and of the outbursts of lava under the surface of great oceanic spaces.

Evidence of great upheavals has been brought forth from the islands of the Arctic Ocean and the tundras of Siberia; from the soil of Alaska; from Spitsbergen and Greenland; from the caves of England, the forest-bed of Norfolk, and the rock fissures of Wales and Cornwall; from the rocks of France, the Alps and Juras, and from Gibraltar and Sicily; from the Sahara and the Rift of Africa; from Arabia and its harras, the Kashmir slopes of the Himalayas, and the Siwalik Hills; from the Irrawaddy in Burma and from the Tientsin and Choukoutien deposits in China; from the Andes and the Altiplano; from the asphalt pits of California; from the Rocky Mountains and the Columbia Plateau; from the Cumberland cave in Maryland and Agate Spring Quarry in Nebraska; from the hills of Michigan and Vermont with skeletons of whales on them; from the Carolina coast; from the submerged coasts and the bottom of the Atlantic with its Ridge, and the lava bottom of the Pacific.

[4] Pettersson, *Scientific American*, August 1950.
[5] Pettersson, *Westward Ho with the Albatross*, p. 150.

With many other places in various parts of the world we shall deal in some detail in the pages that follow; but we shall not exhaust the list, for there is not a meridian of longitude or a degree of latitude that does not show scars of repeated upheavals.

Chapter VIII

POLES DISPLACED

The Cause of the Ice Ages

One after the other, scenes of upheaval and devastation have presented themselves to explorers, and almost every new cave opened, mountain thrust explored, undersea canyon investigated, has consistently disclosed the same picture of violence and desolation. Under the weight of this evidence two great theories of the nineteenth century have become more and more strained: the theory of uniformity and the theory of evolution built upon it. The other fundamental teaching originating in the nineteenth century—the theory of ice ages— has been loaded more and more heavily with the responsibility for the geological facts revealed; however, the cause of the ice ages remained a much-discussed and never agreed-upon subject.

The origin of the glacial periods was sought "on the earth below and in the heaven above." The theories that endeavoured to explain what caused them fall under the following headings: astronomical, geological, and atmospherical.

In the first group, some theories seek the cause of the ice ages in space, some in the sun, some in the relative positions of the sun and the earth. One idea was that the space through which the solar system travelled was not always of equally low temperature, the variations being due to gases or dust present in some areas. This idea has been abandoned. Another theory was that the sun is a variable star emitting more heat at some periods and less at others. This theory also failed to be substantiated and was generally rejected; yet sporadically it finds new proponents.[1] Still another theory would have the ice ages arrive

[1] Barbara Bell, *Science Newsletter*, May 24, 1952.

when a hemisphere, the Northern or the Southern, happens to have its winter while the globe is at the farthest end of its ellipse, as the Southern Hemisphere is at present. The winter would be a little longer and colder; however, the summer, though a bit shorter, would be hotter, and if the earth always travelled on its present orbit, the described variations would not bring about an ice age. It was also claimed that the terrestrial orbit becomes alternately more and less stretched.

Of the geological group of theories, one supposed a change in the activity of warm springs; another, a change in the direction of the Gulf Stream, which carries water warmed in the Caribbean Sea to the northern Atlantic; if there were no Isthmus of Panama, and North and South America were separated, a part of the stream from the Caribbean would flow into the Pacific. Both these theories were shown to be inadequate, and the paleontological survey of sea fauna on both sides of the isthmus suggests that the dividing strip of land existed long before the advent of the Ice Age. Another geological theory, which still has some adherents, sees the origin of the glacial periods in the changing altitude of the continents, which would also influence the direction of winds and precipitation. But it is definitely opposed by such an authority on glacial geology as A. P. Coleman, professor emeritus of geology at Toronto University:

"When one considers the distribution of ice sheets in the Pleistocene, covering 4,000,000 square miles of North America and half as much of Europe . . . [and the ice in] Greenland, Iceland, Spitsbergen . . . the southern island of New Zealand and Patagonia in South America, it becomes evident that all parts of the world could not have been elevated at once. The theory breaks down of its own weight." Therefore "elevation above snow line would cause local glaciation, but there is no evidence that large scale ice sheets can be formed in this way, and that a universal refrigeration, like that of the Pleistocene, could be produced thus is manifestly impossible."[2]

Of the atmospheric conditions that could effect a rise or drop in temperature, the varying quantity of carbon dioxide in the air and also of dust particles was called on to explain the changes in temperature in the past. With the diminution of the

[2] A. P. Coleman, *Ice Ages Recent and Ancient* (1926), p. 256.

97

carbon dioxide content in the air there would be a fall in the temperature, but it was demonstrated by calculation that this could not have been sufficient to cause the Ice Age. If the earth were enveloped in clouds of dust that kept the rays of the sun from penetrating to the ground, there would be a fall in temperature. However, one would have to explain where such extensive and thick clouds of dust in the atmosphere could come from.

"Scores of methods of accounting for ice ages have been proposed, and probably no other geological problem has been so earnestly discussed, not only by geologists but by meteorologists and biologists; and yet no theory is generally accepted."[3]

A true theory of the origin of ice ages, whether resorting to astronomical, geological, or atmospheric causes, must also explain why ice ages did not occur in northeastern Siberia, the coldest place on earth, but did occur in temperate latitudes, and in a much more remote past in India, Madagascar, and equatorial Brazil. None of the theories mentioned explains these strange facts. Hypotheses concerning warmer and colder areas in space, or the variability of the sun as a source of energy, are especially inadequate to account for the geographical distribution of the ice cover. Thus the concept of ice ages, which is established in science as one of its most definite facts, serving also as a foundation for the theory of evolution, has no explanation itself.

Shifting Poles

All other theories of the origin of the Ice Age having failed, there remained an avenue of approach which already early in the discussion was chosen by several geologists: a shift of the terrestrial poles. If for some reason the poles had moved from their original positions, old polar ice would have moved out of the Arctic and Antarctic circles and into new regions. The glacial cover of the Ice Age could have been the polar icecap of an earlier epoch. Thus would be explained not only the origin of the ice cover but also the fact that its geographical position did not coincide with the present Polar Circles.

[3] Ibid., p. 246.

"The simplest and most obvious explanation of great secular changes in climate, and of former prevalence of higher temperatures in northern circumpolar regions, would be found in the assumption that the earth's axis of rotation has not always had the same position, but that it may have changed its position as a result of geological processes, such as extended rearrangement of land and water."[1]

For decades in the second half of the nineteenth century many scientists participated in the debate centring round this theme. Astronomers and mathematicians asked the geologists what, in their opinion, could have caused such a shifting of the terrestrial poles. The best the geologists could offer was the redisposition of the weight on the surface of the earth. Sir George B. Airy, astronomer royal, analysed the question by assuming that the earth, a perfectly rigid spheroid (a flattened globe), was disturbed in its rotation by a sudden elevation of a mountainous mass in latitudes "most favourable for production of a large effect." The axis of rotation would no longer coincide with the axis of figure, and there would be wobbling. "Under these circumstances, the axis of rotation would wander in the solid earth. But it would not wander indefinitely. . . ."

However, the smallness of the effect would be disappointing. If a mountain mass should be produced equal to one one-thousandth part of the mass of the equatorial bulge—"which I apprehend is very far above the fact . . . the shift of the earth's pole would be only two or three miles, and this, though it would greatly surprise astronomers . . . would produce no such changes of climate as those which it is desired to explain."[2]

Sir George Darwin, mathematician and cosmologist, the renowned son of an illustrious father, made more thorough calculations on this point. If an ocean bed 15,000 feet deep rose to become a continent the size of Africa 1100 feet above sea level, and on the other side of the globe an equal area became depressed, the effect would be a shift of the poles by about two degrees. However, were the earth plastic, the poles would wander to a greater extent.

James Croll, the Scottish climatologist, wrote:

[1] Julius Hann (Austrian meteorologist, 1839–1921), quoted by W. B. Wright, *The Quaternary Ice Age*, p. 313.
[2] *Athenaeum*, September 22, 1860, p. 384.

"There probably never was an upheaval of such magnitude in the history of our earth. And to produce a deflection of 3° 17′ —a deflection that would hardly sensibly affect climate—no less than one-tenth of the entire surface of the earth would require to be elevated to the height of 10,000 feet. A continent ten times the size of Europe elevated two miles would do little more than bring London to the latitude of Edinburgh, or Edinburgh to the latitude of London. He must be a sanguine geologist indeed who can expect to account for the glaciation of this country, or for the former absence of ice around the poles, by this means. We know perfectly well that since the Glacial epoch there have been no changes in the physical geography of the earth sufficient to deflect the pole half a dozen miles, far less half a dozen degrees."[3]

J. Evans, a geologist, suggested that the astronomers reconsider their conclusion, on the supposition that the earth is a shell filled with molten matter. He envisaged the possibility that, under a change of load in the crust, the crust would be forced to alter its position in relation to the axis by as much as twenty degrees.[4]

Sir William Thomson (Lord Kelvin), the physicist, took up the issue and retorted that "the earth cannot, as many geologists suppose, be a liquid mass enclosed in only a thin shell of solidified matter."[5] "At the surface and for many miles below the surface, the rigidity [of the earth] is certainly very much less than that of iron; and therefore at great depths the rigidity must be enormously greater than at the surface. . . . Whatever be its age, we may be quite sure the earth is solid in its interior . . . and we must utterly reject any geological hypothesis which . . . assumes the solid earth to be a shell of 30 or 100, or 500 or 1000 kilometres thickness, resting on an interior liquid mass."

Lord Kelvin showed that, if the earth were a liquid mass covered with a solid crust, "the solid crust would yield so freely to the deforming influence of sun and moon, that it would simply carry the waters of the oceans up and down with it, and there would be no sensible tidal rise and fall of water relatively to it. The state of the case is shortly this: The hypothesis of a

---

[3] J. Croll, *Discussions on Climate and Cosmology* (1886), p. 5.
[4] J. Evans, *Journal of the Geological Society of London*, XXXIV, 41.
[5] Thomson, *British Association for the Advancement of Science, Report of the 46th Meeting, 1876, Notices and Abstracts* (1877), pp. 6, 7.

perfectly rigid crust containing liquid, violates physics by assuming preternaturally rigid matter, and violates dynamical astronomy. . . ."[6]

Lord Kelvin admitted, however, that a larger shifting of the poles would be possible if the earth had a solid nucleus in the interior separated by a liquid layer from the outer crust. This he regarded as improbable and directed his argument against an earth with a molten interior.

George Darwin supported the views of Lord Kelvin, presenting figures to show that the earth could not have a fluid nucleus; its rigidity must be at least as great as that of steel.[7]

Thus the efforts of the geologists to explain the origin of the ice cover by the shifting of the poles foundered on the calculations of the mathematicians. A mathematician made the point clear:

"Mathematicians may seem to geologists almost churlish in their unwillingness to admit a change in the earth's axis. Geologists scarcely know how much is involved in what they ask. They do not seem to realize the vastness of the earth's size, or the enormous quantity of her motion. When a mass of matter is in rotation about an axis, it cannot be made to rotate about a new one except by external force. Internal changes cannot alter the axis, only the distribution of the matter and motion about it. If the mass began to revolve about a new axis, every particle would begin to move in a different direction. What is to cause this? . . . Where is the force that could deflect every portion of it, and every particle of the earth into a new direction of motion?"[8]

Searching for causes in the earth itself, the geologists offered a theory concerning changes on the surface of the globe which, as the astronomers calculated, could have displaced the poles— but only to an extent entirely inadequate to account for the ice cover in the Ice Age. The explanation that appeared best to the geologists was rejected by the physicists and astronomers, who for their part could not propose any other satisfactory solution.

[6] Ibid.
[7] George Darwin, "A Numerical Estimate of the Rigidity of the Earth," *Nature*, XXVII (1882), 23.
[8] *Geological Magazine* (1878), 265.

Further developments showed that tides in the terrestrial crust under the influence of moon and sun, unknown to Lord Kelvin, do exist, though they are minute; this means that the earth is not perfectly rigid. It was also found that the earth makes a real wobbling motion. S. C. Chandler, an American astronomer (1846–1913), explained the wobbling of the earth as an indication of its being removed from a balanced position. Simon Newcomb, foremost American mathematical astronomer, in his paper, "On the Periodic Variation of Latitude," wrote:

"Chandler's remarkable discovery that the apparent variation in terrestrial latitudes may be accounted for by supposing a revolution of the axis of rotation of the earth around that of figure . . . is in such disaccord with the received theory of the earth's rotation, that, at first, I was disposed to doubt its possibility." However, on reconsideration, he found a theoretical justification: "Theory then shows that the axis of rotation will revolve around that of figure, in a period of 306 days and in a direction from west toward east."[9]

G. V. Schiaparelli, the Italian astronomer, in his research, *De la rotation de la terre sous l'influence des actions géologiques* (1889), pointed out that in the case of displacement the pole of inertia (or of figure) and the new pole of rotation would describe circles around each other, and the earth would be in a state of strain. "The earth is at present in this condition and as a result the pole of rotation describes a small circle in 304 days, known as the Eulerian circle."[10] This phenomenon of wobbling points to a displacement of the terrestrial poles sometime in the past. The question centres, then, on the forces that could have caused such a shift.

Schiaparelli wrote: "The performance of the geographical poles in the very same regions of the earth cannot yet be considered as incontestably established by astronomical or mechanical arguments. Such permanence may be a fact today, but it remains a matter still to be proven for the preceding ages of the history of the globe."[11] He thought that a series of geological changes could, by their cumulative effect, step by step, destroy

[9] Simon Newcomb, *Astronomical Journal*, XI (1891); Cf. idem, in *Monthly Notices of the Royal Astronomical Society*, LII (1892), No. 35.

[10] Later observations put the Eulerian, or Chandler's, period at 428 to 429 days.

[11] G. V. Schiaparelli, *De la rotation de la terre*, p. 31.

the equilibrium of the earth, on condition that the earth is not an absolutely rigid body. "The possibility of great shifting of the pole is an important element in the discussion of prehistoric climates and the distribution, geographic and chronologic, of ancient organisms. If this possibility is admitted, it will open new horizons for the study of great mechanical revolutions that the crust of the earth underwent in the past. We cannot imagine, for instance, that the terrestrial equator could take the place of a meridian, without great horizontal tension in some regions, that would open great rifts; and in other regions, horizontal compressions would have taken place, such as are imagined today in order to explain the folding of the strata and formation of mountains."

The resistance of the spheroid or the terrestrial globe, flattened at the poles, to a change in position must, in Schiaparelli's opinion, show itself also in the levelling of great areas and the extension of shallow seas, like that of the Baltic and North seas. He finished by saying: "Our problem, so important from the astronomical and mathematical standpoint, touches the foundations of geology and paleontology; its solution is tied to the most grandiose events in the history of the Earth."

Thus, finally, an eminent astronomer, after a thorough examination of the problem, went over to the side of the geologists. However, he reasoned in a circle: the geological changes would cause the terrestrial poles to move from their places, and the motion of the poles from their places would cause geological and climatic changes.

A gradual and slow displacement of the poles or a tilting of the axis would explain the geographical position of the ice in the past, but it would not account for other phenomena observed, such as the extent of the glacial cover and the suddenness with which it enveloped the earth. Agassiz realized this, and in support of the idea that the ice ages came suddenly, he quoted Cuvier. Cuvier died before the Ice Age theory was promulgated, but he understood that the climate must have altered suddenly in order to encase the large quadrupeds of Siberia in ice as soon as they were killed, and to preserve their bodies from decay since then. "Therefore," wrote Cuvier in prophetic anticipation of the debate that has been renewed for over a hundred years, down to our times, "all hypotheses of a *gradual* cooling of the earth, or a *slow* variation of the inclination or position of the terrestrial axis, are inadequate."[12]

The Sliding Continents

The geological changes in the distribution of land and water being inadequate to explain the shifting of the poles, the problem is thrown back once more into the domain of astronomy. But before we ask. "What forces in the solar system could have displaced the terrestrial axis?" we shall discuss a theory that has for over three decades occupied the minds of geologists, climatologists, and evolutionists—the theory of the shifting continents. Instead of the poles shifting, according to Wegener's theory the continents drift and pass one after the other through the southern and northern polar regions.

In August 1950 the British Association for the Advancement of Science devoted the sessions of its annual convention to debate on the question: Is the theory of the continental drift (slide) right or wrong? There were many defenders of the theory and as many opponents. The theory was then put to a vote. The result was an even division between "yea" and "nay." The chairman was entitled to cast the deciding vote but abstained. Only through the fortuitous circumstance that the presiding officer was a conscientious—or undecided—person was the sanctification of continental drift averted.

The theory of drifting continents, debated since the 1920s, has its starting point in the "similarity of the shapes of the coastlines of Brazil and Africa."[1] This similarity (or, better, complementation), plus some early faunal and floral affinities suggested to Professor Alfred Wegener of Graz in the Tyrol that in an early geological age these two continents, South America and Africa, were one land mass. But since animal and vegetable affinities could also be found in other parts of the world, Wegener conjectured that all continents and islands were once a single land mass that in various epochs divided and drifted apart. Those who do not subscribe to the theory of continental drift continue to explain the affinity of plants and animals by "land bridges"

[12] Agassiz, *Etudes sur les glaciers*, p. 311; Cuvier, *Recherches sur les ossements fossiles* (2nd ed), I, 202.
[1] A. Wegener, *The Origin of Continents and Oceans* (1924), p. 1.

or former land connections between continents and also between continents and islands.

In order that continents might move, it is claimed that there must be a basic difference between the composition of the earth's crust that is exposed in land masses and that which exists on the bottom of the ocean. The theory of drifting continents is grounded on the "increasingly well-proven doctrine of isostasy or the flotation of the crust of the earth" on plastic magma. A new nomenclature was introduced. The land masses or the outer crust are called *sial*, an abbreviation of silicon and aluminium, two of the elements predominant in the composition of terrestrial rocks. The substratum is called *sima*, an abbreviation of silicon and magnesium, there being a "good reason for believing that the rocks forming the substratum [bottom] of the ocean bed are more basic in composition and contain a large proportion of magnesia [magnesium oxide]."[2] It is also assumed that the sima underlies the sial of the continents and, possessing the plastic properties of sealing wax, permits the continents to drift.

Besides accounting for the correspondence between the coastal features of eastern South America and western Africa and between those of other continents, and certain affinities in the animal and plant kingdoms, the theory of drift tries to account for several geological phenomena, all in need of explanation: (1) the cause of the ice ages; (2) the distribution of coal beds; and (3) the formation of mountains. According to Wegener, mountainous crests rose, in the movement of the land, on the forward side of the floating continents; meeting some resistance in íts motion on elastic sima, the sial formed elevations. Thus, when South America moved away from Africa, an elevation was raised on the side turned to the Pacific Ocean, the Andes.

If, from the beginning, there was but one land mass, there could have been only one ocean, too, and, according to Wegener, the only ocean was the Pacific. The Atlantic is a later formation, and its bottom cannot be of sima, like that of the Pacific, but is built of stretched sial. Sufficient proof of the difference in composition of the substrata of the Atlantic and the Pacific has not yet been adduced.

[2] John W. Evans, president of the Geological Society, in Introduction to Wegener, *The Origin of Continents and Oceans.*

The occurrence in an early glacial period of ice cover in lands now in tropical and subtropical regions is explained by the supposition that these lands were once in the Antarctic. However, their extent is so great that if all of them were joined around the South Pole, many parts that have signs of the Ice Age would still be too remote from the pole. The theory assumes, therefore, that these lands occupied in succession the position of the Antarctic continent today, each in its turn passing through a glacial period; the signs of glaciation in Africa, India, Australia and South America are accounted for by the successive sliding of these continents through the southern polar region. A similar explanation is offered for the origin of the Ice Age in the Northern Hemisphere, at a much more recent date, when the land masses of North America and Europe wandered close to the North Pole. The North Pole is charted on various points on the globe—in the Pacific, in the Canadian arctic archipelago, in Greenland, in Spitsbergen—all in succession during the Pleistocene, or recent Ice Age.

The coal beds in northern countries, among them Alaska and Spitsbergen, are dated by Wegener from the time when these lands occupied tropical or subtropical belts, on their passage from the Southern Hemisphere to the Northern.

If a theory can explain the origin of mountains, the cause of the ice ages, the coal beds in higher latitudes, and certain common characteristics of the fauna and flora of continents separated by oceans, then the correspondence in the contours of the Brazilian and West African coasts truly served as a clue to the solution of major problems in geology and climatology. However, there are facts that strongly challenge this hypothesis.

The minute difference between the gravitational pulls to which the crust is subjected in higher latitudes and closer to the equator was offered by Wegener as the motive force in the drift of continents. But Harold Jeffreys, a British cosmologist, computed that this force is one hundred billion times too weak to produce the effect. "There is therefore not the slightest reason to believe that bodily displacements of continents through the lithosphere [the crust] are possible."[3] Even assuming that this motive force was sufficient, why did the lands

[3] H. Jeffreys, *The Earth, Its Origin, History and Physical Constitution* (2nd ed.; 1929), p. 304

of Europe, Siberia, and North America first move away from the original common land mass toward the equator and then retreat from the equator?

In search of another moving force, A. L. du Toit, a South African scientist, offered a variation of Wegener's theory, namely, a "concept of an earth in which the periodic, though variable, softening of the sub-crust through radioactive heating enables the skin to creep differentially over the core with consequent wrinkling."[4]

As for the mountains, not all of them are situated as long ridges parallel to the seacoast. And no compelling evidence has been brought to support the contention that ice ages were consecutive in various parts of the Southern Hemisphere and, in much more recent times, in various parts of the Northern Hemisphere. Furthermore, how explain signs of the recent Ice Age in the Southern Hemisphere? In Patagonia, New Zealand, and other places in the Southern Hemisphere, signs of recent glaciation are found. It is also certain that the chilling of the Ice Age was simultaneous all over the world.

Coal is found not only in arctic lands but also in Antarctica. Did, then, this continent travel there from the tropics? And what was the motive force?

If the theory is correct, the motion of the continents should be observable at present; yet, though Wegener claimed, on the basis of certain reports, that Greenland and an island near its western coast still move, repeated observations and triangulations do not support this claim. Wegener perished on an expedition to Greenland in 1930.

The assumption that ocean floors and continents are eternally different in structure is in contradiction to a great number of observations, though the land surface has been better explored than the bottom of the sea. The idea of a basic difference between the rocks of the ocean bottom and those of the continents is disproved wherever the fossiliferous contents of the land and of the ocean bed are examined. Marine expeditions have failed to find at various places on the ocean bottom the thick layers of sediment that should have been present if the sea had been covering the areas for untold centuries. On the other hand, sediments thousands and even tens of thousands of feet

[4] A. L. du Toit, *Our Wandering Continents* (1937), p. 3.

thick have been found on continents. Not only were large stretches of land in North America and Europe and Asia covered by the sea at various times in the past—and some well-investigated localities, like the gypsum beds of Paris, show repeated returns of the waters—but even the largest and highest mountain chains—the Alps, the Andes, the Himalayas—at some time were under the sea. Since the ocean once covered a vast expanse of land, it may at present occupy the place of former land.

The land masses of today do not change their latitudes; the motive force claimed is insufficient by far. Coal beds in Antarctica and recent glaciation in temperate latitudes of the Southern Hemisphere all conspire to invalidate the theory of wandering continents.

The Changing Orbit

The theory of sliding continents having been shown to be built on infirm foundations, there remain three theoretical changes in the position of the terrestrial globe or its shell in relation to the sun that could cause great variations of climate: a change in the form of the orbit, or the path the earth follows around the sun; a change in the astronomical direction of the axis; a change in the position of the terrestrial shell in relation to the core, and thus in the position of the poles (sliding shell).

At present the elliptical form of the orbit changes by a very small amount. This could be the residue of a displacement the earth suffered on its path; but following the principle of Laplace and Lagrange concerning the stability of the planetary system, this variation in the shape of the terrestrial orbit is considered to be an oscillation, the mean shape of the orbit being regarded as fixed. The period of this oscillation is supposed to be of very long duration.

The obliquity of the ecliptic, or the angle which the plane of the equator makes with the plane of the earth's orbit, is $23\frac{1}{2}°$; this obliquity causes the sequence of the seasons. It changes now at the rate of 0.47″ a year, "but the limits of its variation are difficult to calculate."[1] The figures offered by various mathema-

[1] Brooks, *Climate through the Ages* (2nd ed.; 1949), p. 102.

ticians differ greatly. Lagrange estimated the angle of the swing to be as large as 7° with a period that had its last maximum in the year 2167 before the present era; Stockwell calculated the angle of oscillation at less than 3°; while Drayson estimated that the obliquity ranged from 35° to 11°, that is, a swing of 24°.[2] This variation, whatever its numerical value, could have been caused by a disturbance which the earth suffered; but again, the cause being unidentified, the effect is considered to be a permanent oscillation.

The earth experiences the precision of the equinoxes, or a large spin of the axis with consequent displacement of the seasons in relation to the perihelion (the point on the orbit closest to the sun). This precession or "preceding" of the vernal and autumnal equinoxes is as great as 50.2″ in a year, and the terrestrial axis describes a wide circle in the sky in a period estimated at about 26,000 years. Newton explained this phenomenon, known since the days of Hipparchus (120 B.C.), as produced by the attractive effect of the sun and the moon on the bulging part of the equator. But this explanation does not account for what in the first place caused the earth's bulging part or equator to take the position under an angle to the plane of terrestrial revolution, or ecliptic.

This swing of the terrestrial axis—as though the globe were a top disturbed in its motion—could also be caused by a disturbance in the motion of the earth experienced sometimes in the past.

Finally, we have already spoken of the wobbling of the terrestrial axis, or its describing a small circle around the geographical pole, or, better, of the wandering of the pole that causes small variations in latitudes, discovered late in the nineteenth century.

A theory that employed the changes in eccentricity of the orbit and the precession of the equinoxes to explain the variations of climate was advanced in 1864 by James Croll, and accepted by Charles Darwin and others; it has since been abandoned, for it requires alternate glacial ages in the Northern and Southern hemispheres, and the evidence contradicts such an order of events.

More recently, M. Milankovitch introduced the third variable, the obliquity of the ecliptic, to correct some of the defects

[2] Ibid.

of Croll's theory. In the opinion of his critics, however, his curve of climatic changes widely upsets geological dates; nor do his variables offer sufficiently effective reasons for the vigorous changes of climate. Besides, he assigned an arbitrary length to the oscillation period of obliquity. And why were there no ice ages during long periods in the past, if the process recurs at calculable intervals?

Thus the inquiry turned once more to a more radical change—the displacement of the terrestrial crust in relation to the core.

The Rotating Crust

The theory that the terrestrial crust is swimming on magma was first offered when J. H. Pratt, in the 1850s, found that the Himalayas, the largest massif on earth, do not exert the expected gravitational pull and do not deflect a plumbline. The astronomer G. B. Airy was surprised, to the point of disbelief in fact; but then he offered a theory that the granite crust, much lighter than the magma underneath, is only sixty miles thick, and that under the mountains, on the inside of the crust, there are reversed mountains, immersed in the heavier magma, which would account for the lack of gravitational pull by mountains.[1] This is the theory of isostasy.

To the study of isostasy and its anomalies (gravitation is, strangely, stronger over deep seas), F. A. Vening Meinesz, Dutch geophysicist and explorer of oceans, made many important contributions. He found in the very structure of the terrestrial crust signs of some violent displacements on a global scale. Thus it is not merely in order to explain the climates of the past that the dislocation of the crust is postulated. In 1943, Vening Meinesz analyzed "the stresses brought about by a change in position of the rigid Earth's crust with regard to the axis of rotation of the Earth." In this analysis he surmised the crust "to have the same thickness everywhere and to behave as an

[1] J. H. Pratt, "On the Attraction of the Himalaya Mountains . . . upon the Plumbline in India," *Philosophical Transactions of the Royal Society of London*, Vol. CXLV (London, 1855). G B. Airy, "On the Computation of the Effect of the Attraction of Mountain-Masses," ibid.

elastic body." He pointed out that if we assume that the crust happened to move clockwise in relation to the core by over 70° the expected effect "shows a remarkable correlation to many major topographic features and also to the shearing patterns of large parts of the Earth's surface, as, e.g., the North and South Atlantic, the Indian Ocean and the Gulf of Aden, Africa, the Pacific, etc. If the correlation is not fortuitous, and this does not appear probable, we have to suppose that the Earth's crust at some moment of its history has indeed shifted with regard to the Earth's poles and that the crust has undergone a corresponding block-shearing."[2]

However, according to the theory of isostasy, the crust is not of the same thickness everywhere, the crustal protuberances are immersed in a very thick and viscous magma, and for the crust to move, even if it is only sixty miles thick, would require a greater force than is available under prevailing conditions in the solar system or on the earth itself.

The very idea of a crust changing its position in relation to the axis of the interior, or of the globe itself, presupposes the validity of the theory of isostasy. This theory, though generally accepted, finds difficulty in explaining the propagation of seismic waves around the globe.[3] If the earth's crust is not just sixty miles thick—which, in relation to the volume of the magma, is as the thickness of the shell to the content of an egg—but two thousand miles, as some scientists assume, then, of course, the displacement of the crust requires forces nearly as powerful as would the displacement of the entire globe, by inclining its axis into a new position in respect to the cardinal points of the sky.

"We are fully justified in concluding that the lithosphere was displaced during the great Ice Ages, and that the displacements were the direct cause of the alterations in climates during these periods."[4] The author of these lines, K. A. Pauly, propagates the idea offered, or revived, by the astronomer A. E. Eddington

[2] F. A. Vening Meinesz, "Spanningen in de aardrost tengevolge van poolverschuivingen," in *Nederlandsche Akademie van Wetenschappen Verslagen*, Vol. LII, No. 5. (1943).

[3] W. Bowie, "Isostasy," in *Physics of the Earth*, ed. B. Gutenberg (1939), II, 104.

[4] K. A. Pauly, "The Cause of the Great Ice Ages," *Scientific Monthly*, August 1952.

in his paper, "The Borderland of Geology and Astronomy." According to Eddington, the ice ages were caused by the shifting of the earth's outer crust over its interior as a result of tidal friction or the inequality of lunar pull on various layers of the earth; this theory abandons every effort to find in the earth itself the force that might cause the crust in its entirety to change its position in relation to the terrestrial axis, which, in this theory, maintains its astronomical direction. In order to pull the lithosphere, or crust, over the substratum, or core, a lesser force is required than that needed to incline the axis of the whole globe in some new direction, for the crust is but a portion of the entire mass of the earth, and the momentum is dependent on the mass. However, in order to move the crust, preserving the axis of the core as that of the entire globe, the friction between the crust and the substratum must be overcome; and because of the equatorial bulge, in order to alter the position of the crust, it must be stretched in some parts. This would require the application of a great force, which does not appear to exist in tidal friction originating in the moon.

Furthermore, the tidal force acts on the surface of the earth in an east-west direction; and a change in this direction would not disturb the position of the latitude in relation to the pole and could not have been the cause of the ice ages. Eddington's theory requires the sliding of the crust northward and southward; to explain the origin of such sliding, he suggested that the crust, moving slowly in the east-west direction, upon meeting some excessive local friction between itself and the substratum, would change its course. But, as indicated above, the tidal friction of the moon could hardly stretch the crust over the equatorial bulge.

The theory of the sliding lithosphere shares the quantitative inadequacy of the theory of sliding continents. Some motive agent more powerful than tidal friction (Eddington), or gravitational differences at various latitudes (Wegener), or intermittent radioactivity in the earth (Du Toit), must have been at work in order to move continents or the entire lithosphere. Thus these theories meet the fate of the earlier theory that postulated the shifting of the poles because of a geological redistribution of land and sea.

Also the theory that would explain the displacement of the

crust by an asymmetric growth of the polar icecaps, is quantitatively indefensible; this theory uses the same phenomenon—the growing icecaps—as the cause *and* the effect of ice ages.

The present survey of theories, which are quantitatively inadequate yet based on the well-reasoned principle of a change of latitudes or the direction of the axis as the cause of the ice ages, was here undertaken to make clear that thoughtful researchers among geologists, climatologists, and astronomers were unsatisfied with views that would not solve the problem of the geographical distribution of the ice cover in the past, a point of which almost all other theories are strangely oblivious. It follows, then, that the clamour heard at the publication of *Worlds in Collision*, even from some astronomers and geologists, to the effect that the shifting axis or changing latitudes had never been heard of, is not supported by scientific literature.

W. B. Wright, of the Geological Survey of Great Britain, finds that the only way to explain ice ages is to assume that "the earth's axis of rotation has not always had the same position"; and "since it has now become obvious that geological history has witnessed many changes in the position of the climatic zones on the surface of the earth and that at least one notable glaciation, that of the Permo-Carboniferous [preceding the time of the large reptiles], was due to a displacement of the pole from its present position, it becomes worth while to inquire whether the Quaternary [Recent] glaciation would not have a similar cause."[5]

But every inquiry in this direction, in Wright's opinion, failed to find a cause that would account for recurrent but not periodic ice ages; they did not return through geological history at measured intervals. Therefore he concluded: "Among the theories that have been brought forward to account for the phenomena of the Ice Age, there is not a single one which meets the facts of the case in such a manner as to inspire confidence."[6]

Not only must the cause have been more powerful than the agents invoked, but it must have acted with great suddenness. On this we shall dwell in the following sections.

Sudden the agent must have been, and violent; recurrent it must have been, but at highly erratic intervals; and it must have been of titanic power.

[5] Wright, *The Quaternary Ice Age*, p. 313.   [6] Ibid., p. 463.

Chapter IX

AXIS SHIFTED

Earth in a Vice

The displacement of the shell alone requires forces not in existence on the earth itself; and the turning of the earth's axis in a new direction requires more powerful forces still. Of course one change does not preclude the other. Each would result in climatic revolution. If the crust moved, the latitudes would be displaced and, in an extreme case, the poles and the equator could change places; and if the axis turned in a new direction, seasons would change their order and intensity and, in an extreme case, a polar region could be turned for a large part of the year into the warmest place on the globe, being day and night under the direct rays of the sun, as presently is the case with Uranus.

Harold Jeffreys asks in his book *The Earth*: "Has the inclination of the earth's axis to the plane of its orbit varied during its history?" and proceeds: "The answer to [this] question is a definite 'Yes!' The theory of tidal friction . . . assumes the equator and the plane of the earth's and moon's orbits to coincide. The fact [is] that they do not. . . ."[1]

The moon, it is assumed, issued from the equatorial region of the earth by the process of disruption and must, therefore, revolve in the plane of the terrestrial equator; but since it does not, there must have been a displacement either of the moon or of the terrestrial axis; and the position of the moon close to the plane of the ecliptic suggests that the terrestrial axis suffered displacement. Also, if from the beginning there was a difference in the direction of the axes of terrestrial rotation and lunar

[1] Jeffreys, *The Earth*, p. 303.

revolution, this difference must have disappeared as the result of tidal friction. Jeffreys considered the works of George Darwin, who tried to explain the observed positions by recourse to several additional tidal frictions, but he found a flaw in Darwin's hypothesis.

Any internal changes in the earth would be "not important" for the observed change in the direction of the terrestrial axis. Jeffreys says: "If we consider the axis of the earth's angular momentum, this can change in direction *only through couples acting on the earth from outside.*"

The arguments of astronomers against the idea of the geologists concerning the change in the position of the terrestrial axis were correct only in showing that the terrestrial causes could not effectively displace the axis of the earth; but the very fact of displacement is now claimed because of astronomical considerations and by such an authority in this field as Jeffreys. What could have played the role of couples, or a vice, acting from outside? And, again, was it a gradual change or a sudden displacement?

Evaporating Oceans

If we take into account the area occupied by ice in the glacial epoch, much larger than the area of the present polar ice, we must conclude that the shifting of the poles alone cannot explain the origin of the glacial cover. The expansion of the glacial cover in its various stages is supposed to be known. The usual estimate of its thickness is between six and twelve thousand feet. From these figures the mass of the ice is calculated and the quantity of water necessary to produce it. The water must have come from the oceans; it is estimated that the surface of the oceans must have been at least three hundred feet lower when the ice cover was developed. Some estimates double, triple, quadruple, and even increase sevenfold this figure. But for all the oceans to have evaporated to such an extent, turning many areas of the continental shelf (the sea at the coast to a depth of a hundred fathoms, or six hundred feet) into a desert of sand and shells an enormous amount of heat was necessary.

John Tyndall, a British physicist of the last century, wrote:

"Some eminent men have thought, and some still think, that the reduction of temperature, during the glacial epoch, was due to a temporary diminution of solar radiation; others have thought that, in its motion through space, our system may have traversed regions of low temperature, and that during its passage through these regions, the ancient glaciers were produced. . . . Many of them seem to have overlooked the fact that the enormous extension of glaciers in bygone ages demonstrates, just as rigidly, the operation of heat as well as the action of cold. Cold [alone] will not produce glaciers."[1]

Tyndall then went on to demonstrate the amount of heat necessary to transport water to the polar regions in the form of snow. He calculated that for every pound of vapour produced a quantity of heat is required sufficient to raise five pounds of cast iron to the melting point. Consequently, in order to evaporate the oceans and transform the water into aqueous clouds that would later descend as snow and turn to ice, a quantity of heat was needed that would raise to the melting point a mass of iron five times greater than the mass of the ice. Tyndall argued that the geologists should substitute the hot iron for the cold ice, and they would get an idea of the high temperature immediately preceding the Ice Age and the formation of the glacial cover.

If this is so, then none of the theories offered in explanation of the Ice Age really would account for it. Even if the sun disappeared and the earth lost its heat to cosmic space, there would be no Ice Age: the oceans and all the water would freeze, but there would be no ice formation on land.

The importance of heat in the formation of the ice cover of the Ice Age was stressed even more by another author, an astronomer of our day (D. Menzel, of the Harvard Observatory): "*If* solar variability caused the ice ages, I should prefer to believe that increased warmth brought them on, whereas a diminution of heat caused them to stop."[2]

What could have brought about such a rise in the temperature of the oceans that, all over the globe, they evaporated enough to lower their surface not three, not thirty, but more than three hundred feet? Could the heat have been generated by the decomposition of organic matter in the sediment? It goes

[1] John Tyndall, *Heat Considered as a Mode of Motion* (1883), pp. 191–92.
[2] D. Menzel, *Our Sun* (1950), p. 248.

without saying that this source would have been utterly inadequate. A tremendous heating process must have preceded the formation of the ice cover; and since it is generally maintained that there were at least four glacial periods in the recent [Quaternary] Ice Age, in each of which the ice grew and then retreated in the interglacial stage, the globe, in a recent geological epoch, must have been repeatedly so hot that the portion of the heat the oceans received would have sufficed to turn an immense mountain of iron, five times the mass of the continental ice cover, to a white glow and melt it. According to Tyndall, if this did not happen, there could not have been any ice ages.

Do we know under what circumstances the earth and its oceans would be heated on a stupendous scale?

If we subscribe to the Ice Age theory, we must assume that the terrestrial globe with its oceans was heated as in a furnace, in the age of man, since the Ice Age, together with the Recent, is the age of man. Large stretches of the bottoms of the oceans must have bubbled with lava. But what could have produced this simultaneous activity of subterranean heat over such vast areas?

We cannot imagine any cause or agent for this, unless it be an exogenous agent, an extraterrestrial cause. For the removal of the poles from their places, or the shifting of the axis, also, only an external agent could have been responsible. The adherents of the Ice Age theory must look to the celestial for the causes of at least four separate encounters, in the not too distant past, with some celestial mass of matter or field of force.

On passing through a large cloud of dust particles or meteorites, the earth and its atmosphere would be heated by the direct impact of these bodies on its atmosphere, its oceans, its land. Under such an impact a displacement of the poles or disturbance in axial rotation would also produce heat in every particle of the globe, because of the conversion of a portion of the energy of motion into heat. This is one theoretical possibility.

The other possibility would be that, on passing through a cloud of dust carrying an electromagnetic charge, the earth would react with electrical currents on its surface that would develop a thermal effect. If the earth passed through a strong field, the heat would be very intense. Selecting the better conducting strata, these currents would go through metal-bearing formations, possibly deeper in the crust, sparing life in some quarters and destroying it in others. Such heat could evaporate oceans to a great depth, cause the intrusion of igneous rock into sedimentary rock, start the flow of magma from fissures, and activate all volcanoes.

The earth is itself a large magnet. A charged cloud of dust or gases, moving in relation to the earth, would be an electro-magnet. An extraneous electromagnetic field that would pro-duce a thermal effect on the earth would also shift the terrestrial axis and change the rotational velocity of the earth. This, in turn, would have a thermal effect, since the energy of motion would be converted into heat, and possibly into other forms of energy—electrical, magnetic, and chemical, as well as nuclear —with ensuing radioactivity, again with thermal effect.

An extraneous mechanical or electromagnetic force would produce both phenomena, which are prerequisites of a glacial period: the astronomical or geographical shifting of the axis and the heating of the globe. The astronomers who oppose the theory of cosmic catastrophes must likewise reject the theory of the ice ages.

Condensation

In the preceding section it was made clear that, for the ice cover of the glacial epoch to be formed, evaporation of the oceans on a large scale must have occurred. But evaporation of the oceans would not be enough; rapid and powerful condensa-tion of the vapours must have followed. "We need a condenser so powerful that this vapour, instead of falling in liquid showers to the earth, shall be so far reduced in temperature as to descend as snow."[1]

An unusual sequence of events was necessary: the oceans must have steamed and the vaporized water must have fallen as snow in latitudes of temperate climates. This sequence of heat and cold must have taken place in quick succession.

A precipitate drop in temperature and a rapid condensation of vapours could have followed from the screening effect of clouds of dust. Dust of either volcanic or meteoric origin, by enveloping the earth, could have impaired the solar light and

[1] Tyndall, *Heat Considered as a Mode of Motion*, pp. 188–89.

warmth reaching the lower atmosphere. The dust particles ejected by erupting volcanoes have been observed to float in the sky around the globe for many months. So, after the eruption of Krakatoa in Sunda Strait between Java and Sumatra in 1883, dust particles suspended in the atmosphere continued for over a year to act as a screen around the world that caused the sunsets to be unusually colourful.[2] Dust from many volcanoes could build a screen that would obstruct the solar light. Actually the screening of the earth by clouds of dust of volcanic origin was one of the theories concerning the origin of ice in the glacial epochs; however, like heat alone, cold alone would not have sufficed to produce the continental ice covers.

In the struggle between heat and cold, snow would descend in some parts of the world and torrential rains in others. And, in fact, numerous scientists who conducted their field study in various areas outside the former ice cover came to the conclusion that those areas had experienced periods of torrential rains that were simultaneous with the glacial periods in higher latitudes. Gregory, studying the African continent, observed signs of water action on a great scale at the same time that other areas were being covered by advancing ice.[3] There remained in the Sahara and adjacent regions stream channels "not now occupied by water courses" that obviously carried great quantities of water. "It is believed probable that these streamways were trenched during a pluvial age or pluvial ages" (Flint). In the Pluvial, Lake Victoria in Africa stood over 300 feet above its present level; since that time there was a complete reversal of the river system in the region.[4] Shor Kul, a salt lake in Sinkiang, had its level 350 feet higher than it is today. Lake Bonneville, which occupied parts of Utah, Nevada, and Idaho, and collected pluvial water as well as melt water from the local glaciers in the mountains, stood "more than 1000 feet above the present Great Salt Lake."[5]

[2] Cf. G. J. Symons, ed., *The Eruption of Krakatoa: Report of The Krakatoa Committee of The Royal Society* (1888), pp. 40ff.

[3] British Association for the Advancement of Science, *Report of the 98th Meeting, 1930* (1931), p. 371.

[4] L. S. B. Leakey, "Changes in the Physical Geography of East Africa in Human Times," *The Geographical Journal of the Royal Geographical Society*, vol. LXXXIV (1934).

[5] Flint, *Glacial Geology*, pp. 472, 479.

Although some geologists, on theoretical grounds, would prefer to think that a dry climate prevailed in the world when so much water was concentrated in the ice covers, field geology shows the opposite to have been the case: snow fell in huge masses, and rain cascaded from the sky at the very same time.

A Working Hypothesis

Let us assume, as a working hypothesis, that under the impact of a force or the influence of an agent—and the earth does not travel in an empty universe—the axis of the earth shifted or tilted. At that moment an earthquake would make the globe shudder. Air and water would continue to move through inertia; hurricanes would sweep the earth and the seas would rush over continents, carrying gravel and sand and marine animals, and casting them on the land. Heat would be developed, rocks would melt, volcanoes would erupt, lava would flow from fissures in the ruptured ground and cover vast areas. Mountains would spring up from the plains and would travel and climb on the shoulders of other mountains, causing faults and rifts. Lakes would be tilted and emptied, rivers would change their beds; large land areas with all their inhabitants would slip under the sea. Forests would burn, and the hurricanes and wild seas would wrest them from the ground on which they grew and pile them, branch and root, in huge heaps. Seas would turn into deserts, their waters rolling away.

And if a change in the velocity of the diurnal rotation—slowing it down—should accompany the shifting of the axis, the water confined to the equatorial oceans by centrifugal force would retreat to the poles, and high tides and hurricanes would rush from pole to pole, carrying reindeer and seals to the tropics and desert lions into the Arctic, moving from the equator up to the mountain ridges of the Himalayas and down the African jungles; and crumbled rocks torn from splintering mountains would be scattered over large distances; and herds of animals would be washed from the plains of Siberia. The shifting of the axis would also change the climate of every place, leaving corals in Newfoundland and elephants in Alaska, fig trees in northern Greenland and luxuriant forests in Antarctica. In the event of a rapid shift of the axis, many species and genera of animals on land and in the sea would be destroyed, and civilizations, if any, would be reduced to ruins.

Water evaporated from the oceans would rise in clouds and fall again in torrential rains and snowfalls. Clouds of dust, ejected by numerous volcanoes and swept by hurricanes from the ground, and possibly dust clouds of extraneous origin—if a cometary train of meteorites was the foreign body causing the upheaval—all this dust would keep the rays of the sun from penetrating to the earth. The temperature under the clouds would be reduced, but close to the ground it would be higher than normal because the heated earth would, by convection, dissipate its heat into the atmosphere. Great streams would be formed by the melting ice of the polar regions, carried out of the Polar Circle, and heated by the ground. Glaciers from the mountains would dissolve and inundate the valleys. In higher and in temperate latitudes the falling snow would turn to water or even vapour before reaching the ground or soon thereafter.

For many months and probably years, the snow falling on the ground would melt and run in great streams to the sea, cutting new river channels and carrying off great masses of debris.

Falling again and again in a sunless world, the snow, shielded from the sun's rays by thick clouds enveloping the earth, would finally cool the ground to the point where it would turn, not into water, but into ice. At first this ice would not lie firmly on the ground; from inclines and slopes it would slide down to the deeper valleys and then toward the sea. Large icebergs would fill the sea and, tossing about, melt and drop a load of stones or other detrital material to the bottom; other icebergs, floating over valleys filled with water, would deposit their loads there. In the course of the years the incessant action of the snow would cool the ground in the higher latitudes to such an extent that a permanent cover would be built. And the earth would go on shuddering for centuries, slowly quieting down, and as time passed one after another the volcanoes would burn themselves out.

This catastrophic shifting of the axis, once or a number of times, is presented here only as a working hypothesis but, without exception, all its potential effects have actually taken place.

Assuming now that the working hypothesis is wrong, we are

faced with the necessity of finding a special explanation for each and every phenomenon observed.

The mountains rose from the beds of the seas and folded and faulted. "What generates the enormous forces that bend, break, and mash the rocks in mountain zones? Why have sea floors of remote periods become the lofty highlands of today? These questions still await satisfactory answers."[1]

Climate changed, and the continental ice cover formed. "At present the cause of excessive ice making on the lands remains a baffling mystery, a major question for the future reader of earth's riddles."[2]

Species and genera of animals were extinguished. "The biologist is in despair as he surveys the extinction of so many species and genera in the closing Pleistocene [Ice Age]."[3] Equally sudden and unexplained changes accompanied the close of each geological period.

What caused tropical forests to grow in polar regions? What caused volcanic activity on a great scale in the past and lava flows on land and in the ocean beds? What caused earthquakes to be so numerous and violent in the past? Puzzlement, despair, and frustration are the only answers to each and every one of these phenomena.

The theories of uniformity and evolution maintain that the geological record bears witness that from time immemorial, even from the time this planet began its existence, only minute changes—caused by the wind blowing on the rocks, the sand grains swimming to the sea—accumulated into the vast changes. These causes, however, are inadequate to explain the great revolutions in nature, and they evoke expressions of futility on the part of the specialists, each in his field.

Ice and Tide

Having shown that only global catastrophes could have brought about the building and spreading of ice covers, I shall

[1] C. R. Longwell, A. Knopf and R. F. Flint, *A Textbook of Geology* (1939), p. 405.
[2] Daly, *The Changing World of the Ice Age*, p. 16.
[3] L. C. Eiseley, "The Fire-Drive and the Extinction of the Terminal Pleistocene Fauna," *American Anthropologist.* XLVIII (1946).

now go on to demonstrate that many effects attributed to ice were caused not by it but by onrushing water. The simplicity with which cosmic catastrophes can explain the origin of the continental ice covers should not make us uncritical. The same catastrophes caused great tides to rush over continents. Both phenomena—waves of translation and ice covers—took place.

Tidal waves traversed continents, moving by inertia when the daily rotation of the earth was disturbed; the ocean water also retreated from the equatorial to the polar regions, returning to the equator with the adjustment of the diurnal rotation. These tidal waves, augmented by others produced by the extraneous fields of force, and by tides generated by submarine earthquakes and hurricanes, were the main agents that dispersed erratic boulders, distributed marine sediment over the land, covered the ground with drift. Invasions of the land by the sea, torrential rains, prodigious snowfalls, floods caused by the melting ice cover, and multitudinous icebergs sliding into the sea, all contributed to the readjustment of the mantle of the earth, shifting the sea-floor sand, the disintegrating rock, the lava, the volcanic and meteoric dust and ashes. The arctic lands were denuded and their detachable mantle was washed away; thus was formed the barren stone surface of the Canadian Shield, its soil being carried away as drift.

The erosion and drift, the excavation of lakes and valleys, and their filling in with clay, boulders, and sand have been ascribed to ice that eroded and moved the debris along. The opponents of the Ice Age theory, the last of which is George McCready Price, pointed to the effect of the ice cover in Antarctica on the rocks beneath: ice plays there a protective and not an eroding role; it shields the underlying rock from the erosive action of the elements and especially of the high-velocity winds that blow most of the year in this part of the world. Yet in rapid motion, with many stone fragments and other debris under it, ice could scratch the bedrock and erode and flute the slopes of valleys. But it is doubtful that the weight of the ice would excavate lake basins in cold, hard rock. The ground was heated, lava gushed out of the earth, formations were softened, and oceans, pouring water and stones on rock and lava, made deep impressions in them. When, after the mountainous ice cover was formed, the ground in a new paroxysm gushed lava under the ice, the latter steamed and, subsiding, pressed with a great weight on the softened ground; in this manner, too, ice could excavate beds of lakes and leave other deep marks on the ground it once covered.

Before the Ice Age theory was conceived, drift and erratic boulders were ascribed to the action of great tidal waves. But with the advent of this theory the role of water in the deposition of drift and erratic boulders was denied. "Gigantic waves," wrote J. Geikie, "were supposed to have precipitated upon the land, and then swept madly on over mountain and valley alike, carrying along with them a mighty burden of rocks and stones and rubbish."[1] This view assumed, however, "the former existence of a cause which there was little in nature to warrant." A late opponent of the Ice Age theory, Sir Henry H. Howorth (1843–1923), sought the origin of such tidal waves in a sudden rise of a mountain chain or in an earthquake of the oceanic bottom.[2]

As we have learned on preceding pages, a disturbance in the axial rotation of the earth must have created a displacement of the oceans and their irruption on land; and this very cause—the disturbance in the axial rotation of the earth—must have acted also in order to build the continental ice covers; it also changed the profile of the earth's crust, lifting some mountains and levelling others.

All this created scenes of the utmost complexity. An example is the old but not antiquated description of the Northeastern United States from Maine to Michigan and New Jersey by J. D. Whitney, professor of geology at Harvard (1875–96). In his work, *The Climatic Changes of Later Geological Times* (1882), he wrote about this area as "a region where the Glacial phenomena exhibit the highest degree of complexity. We are beset with difficulties when we attempt to solve the problem presented by the Northern Drift in Northeastern America. . . . Extreme complexity in the direction of the striation; proof of the former presence of the sea over a part of the region, and of fresh water over another extensive portion; enormous accumulations of detrital material evidently deposited by water; occasional peculiar transportation of boulders in a manner not in harmony

[1] J. Geikie, *The Great Ice Age and Its Relation to the Antiquity of Man* (1894), pp. 25–26.
[2] Howorth, *The Glacial Nightmare and the Flood* (1893), p. xx.

with anything we see ice doing at the present time; occurrence of linear accumulations of sand, gravel, and boulders closely resembling the osar [Scandinavian crests of drift] in character; proofs in some parts of the Drift region of the prevalence during the Glacial epoch of a colder climate, and in others of one warmer than that now existing—these are some of the difficulties which must be met by those who undertake to solve the problem of the Northern Drift of Northeastern America."[3] The theories of warm interglacial periods and of the deformation of land and its submersion as the result of the removal of the ice cover could explain the puzzling phenomena in some cases, but in many others they could not do so. Thus bones of seal and walrus are found in Holderness, Yorkshire, with fresh-water molluscs of warm climate. "Despite its anomalous elements, the deposit is classed as interglacial."[4] In similar strata in Yorkshire hippopotami are found, too.

The glaciers in the Alps served as observational material for deductions concerning the continental ice cover. However, alpine glaciers carry stones downhill, not uphill, and the general question was asked whether ice could carry rocks uphill.[5]

Erratic boulders are often found in places where continental ice could hardly have deposited them. Charles Darwin inquired and learned that erratics are found on the Azores, islands separated from the ice cover by a wide expanse of ocean.

Cumming described erratics close to the summit of the Isle of Man in the Irish Sea, where only waves could have lifted them.[6] In Labrador boulders have been seen, rammed against the slopes of the hills, which could have been done only by a tidal wave. As already said, in India, in an earlier ice age, detritus and blocks were carried, not from the land toward the sea, but in the opposite direction, from the sea up the Himalayas, and not from higher latitudes toward lower, but in the opposite direction. The whales in the hills of Vermont and Quebec were cast there by an irrupting ocean.

The very profusion of erratic boulders in many places of the world, sometimes covering wide stretches of a country, whether

[3] J. D. Whitney, *The Climatic Changes of Later Geological Times* (1882), p. 391.
[4] Flint, *Glacial Geology*, p. 342.
[5] G. F. Wright, *The Ice Age in North America*, p. 634,
[6] J. G. Cumming, *Isle of Man*, pp. 176–78.

carried by ice or by tides, presents the problem of their origin: they must have been broken off the mountains in great numbers at a time when ice and water were thrown into action. The mountains must have been under stress, the massifs must have been heated and split, or shattered by earthquakes; they must have been mashed and twisted and rent when the seas trespassed their borders and carried their billows to mountainous ridges, red and bursting.

Magnetic Poles Reversed

When rock is liquefied it is non-magnetic, but, cooling to about 580° Centigrade (Curie point), it acquires a magnetic state and orientation dependent upon the magnetic field of the earth. After solidifying, lava rock retains its magnetic property, and it would retain it even though it became displaced or the magnetic orientation of the earth changed.

In all parts of the globe rock formations are found with reversed polarization;[1] paleomagnetism almost every month detects more areas of inverted orientation. "Sufficient experiments have now been made to allow only one plausible explanation of this 'inverted' magnetization—that the Earth's magnetic field was itself reversed at the period when the rocks were formed."[2] At the same time it was admitted that "no known mechanical or electromagnetic [local] effect can cause a reversal of magnetization over such an area."[3]

An even more puzzling fact is that the rocks with inverted polarity are much more strongly magnetized than can be accounted for by the earth's magnetic field. Lava or igneous rock, on cooling below the Curie point, acquires a magnetic charge stronger than the charge this rock would acquire in the same magnetic field at outdoor temperature, but only doubly so.[4] The rocks with inverted polarity, however, are magnetically charged ten times and often up to a hundred times stronger

---

[1] A. McNish, "On Causes of the Earth's Magnetism and Its Changes," in *Terrestrial Magnetism and Electricity*, ed. J. A. Fleming (1939), p. 326.

[2] H. Manley, "Paleomagnetism," *Science News*, July, 1949, p. 44.

[3] Ibid., pp. 56–57.

[4] The intensity of the acquired magnetic state depends on the velocity with which the lava cools and on the form, size, and composition of its particles.

than they could have been by terrestrial magnetism. "This is one of the most astonishing problems of paleomagnetism, and is not yet fully explained, although the facts are well attested."[5]

Thus we are confronted with an ever growing puzzle. The cause of the reversal of the magnetic field in the rocks of the earth is unknown and the fact contradicts every cosmological theory. The strength of magnetization of the rocks with inverted polarity is astonishing.

Now, if the earth's axis changed its direction or position under the influence of an external magnetic field, we should expect to find the following:

The external magnetic field would create eddy (electrical) currents in the surface layers of the earth; the currents would create a magnetic field around the earth that would counteract the external magnetic field. The strength of the magnetic field created by the eddy currents would be dependent on the external magnetic field and the velocity with which the earth travelled through it. The thermal effect of the electrical currents would liquefy the rocks. The process would be accompanied by volcanic activity and intrusion of igneous rock into surface sedimentary rocks. The molten rock would acquire a magnetic state as soon as its temperature dropped to about 580°C.; also, those rocks that were heated below this temperature would acquire the orientation of the prevailing magnetic field. It is also apparent that an external magnetic field that could shift the terrestrial axis in a short time would have to be of considerable intensity.

We have all three expected effects: lava flowed and igneous rock intruded in the form of dykes or otherwise; the heated rocks acquired a reversed magnetic orientation; the intensity of their magnetization is stronger than the earth's own field could possibly produce.

In the section "A Working Hypothesis," it was asserted that the formation of the ice cover, pluvial phenomena, and mountain building could be explained if the earth's axis was shifted, and it was assumed that the axis was shifted by an extraneous magnetic field. Now, the circumstance that rocks the world over show reversed magnetic orientation and an intensity of magnetization which the earth's magnetic field could not have induced, proves that our assumption was not unfounded.

[5] Ibid., p. 59.

In a recent article, S. K. Runcorn of the University of Cambridge reports that "the evidence accumulates that the earth did reverse its field many times."[6] "The north and south geomagnetic poles reversed places several times . . . the field would suddenly break up and reform with opposite polarity."

The source of the terrestrial magnetism is supposed to be in electrical currents on the surface of the terrestrial core. "Substantial changes in the speed of earth's roatation become easier to explain.

"Whatever the mechanism [of the origin of the terrestrial magnetic field], there seems no doubt that the earth's field is tied up in some way with the rotation of the planet, And this leads to a remarkable finding about the earth's rotation itself."

The unavoidable conclusion, according to Runcorn, is that "the earth's axis of rotation has changed also. In other words, the planet has rolled about, changing the location of its geographical poles." He charted the various positions of the north geographical pole.

The next question, then, is: When was the terrestrial magnetic field reversed for the last time?

Most interesting is the discovery that the last time the reversal of the magnetic field took place was in the eighth century before the present era, or twenty-seven centuries ago. The observation was made on clay fired in kilns by the Etruscans and Greeks.

The position of the ancient vases during firing is known. They were fired in a standing position, as the flow of the glaze testifies. The magnetic inclination or the magnetic dip of the iron particles in the fired clay indicates which was the nearest magnetic pole, the south or the north.

In 1896 Giuseppe Folgheraiter began his careful studies of Attic (Greek) and Etruscan vases of various centuries, starting with the eighth century before the present era. His conclusion was that in the eighth century the earth's magnetic field was inverted in Italy and Greece.[7] Italy and Greece were closer to the south than to the north magnetic pole.

[6] S.K. Runcorn, "The Earth's Magnetism," *Scientific American*, September 1955.
[7] G. Folgheraiter in *Rendi Conti dei Licei*. 1896, 1899; *Archives des sciences physiques et naturelles* (Geneva), 1899; *Journal de physique*, 1899; P. L. Mercanton, "La méthode de Folgheraiter et son rôle en géophysique," *Archives des sciences physiques et naturelles*, 1907.

P. L. Mercanton of Geneva, studying the pots of the Halstatt age from Bavaria (about the year 1000 B.C.) and from the Bronze Age caves in the neighbourhood of Lake Neuchâtel, came to the conclusion that about the tenth century before the present era the direction of the magnetic field differed only a little from its direction today, and yet his material was of an earlier date than the Greek and Etruscan vases examined by Folgheraiter. But checking on the method and the results of Folgheraiter, Mercanton found them perfect.

An ancient vase found by F. A. Forel in Boiron de Morges, on Lake Geneva, was broken and its pieces were scattered and lay in all directions; when assembled, they all showed one and the same magnetic orientation, which proves again that the magnetic field of the earth was unable to change the orientation originally acquired by the clay when fired and cooled in the kiln.[8]

These researches, continued and described in a series of papers by Professor Mercanton, at present with the Service Météorologique Universitaire in Lausanne, show that the magnetic field of the earth, not very different from what it is today, was disturbed sometime during or immediately following the eighth century to the extent of complete reversal.[9]

The eighth and the beginning of the seventh centuries before the present era were periods of great cosmic upheavals, described in *Worlds in Collision*, pages 203–357. At one of the occurrences the solar motion appeared to be reversed, reflecting some disturbance in the terrestrial motion.

Volcanoes, Earthquakes, Comets

A great chain of volcanoes girdles the Pacific Ocean. The Andes in South America are studded with many volcanic summits, among them the loftiest volcanic mountain in the world: Cotopaxi in Ecuador is over 19,000 feet high. The Andes

---

[8] *Bulletin de la Société Vaudoise des sciences naturelles*, Séance du 15 décembre 1909.

[9] Manley speaks of "the possibility of its [earth's magnetic field] reversal in historical times, 2500 years ago, to be cleared up by more research." However, the more exact date is, according to the original works of Folgheraiter and Mercanton, the eighth century before the present era, or shortly thereafter.

reached their present height only in the age of modern man. Magma intruded into the rock and lifted it; in many places magma reached the surface, broke through vents, and built craters. Most of those volcanoes, however, are already extinct.

Central America abounds in volcanoes, most of them extinct or dormant; the highest, Orizaba in Mexico, over 18,000 feet high, was active for the last time three centuries ago. In the United States few volcanoes are active, though many became extinct very recently, in the geological sense. Alaska, the Aleutian Islands, the Kamchatka Peninsula, and the Kurile Islands encircle the northern Pacific with a volcanic arc. The Japanese islands contain volcanoes by the score; most of them are extinct, some only recently so. Formosa, the Philippines, the so-called Volcano Islands—one of which is Iwo Jima—the Moluccas, northern New Zealand, the Sunda Archipelago—all are crowded with volcanoes, most of them only recently extinct. In the centre of this chain are the Hawaiian Islands, with fifteen great mountainous volcanoes, all extinct or dormant except Mauna Loa and Kilauea, two of the largest volcanoes on earth. "How was the 30,000 foot cone built from the floor of the deep sea?"[1] When, in 1855, Mauna Loa erupted, the lava ran over the land at a velocity of forty miles an hour, faster than a swift horse. In 1883, when the volcanic island of Krakatoa in the Sunda Strait blew off, it sent a column of pumice and ashes over seventeen miles high; it raised tides 100 feet high that carried steamships miles inland and were felt on the eastern coast of Africa and the western coast of the Americas as far as Alaska; it created a noise that was heard in Ceylon, in the Philippines, and even in Japan over three thousand miles away. This would compare with an explosion in London heard in New York. When Bandai erupted in Japan in 1888, it cast up almost three billion tons of material and blew off one of its four peaks. But these delayed actions of single volcanoes look like child's play when compared with the forces that in past ages thrust up the Andes, spread the Deccan trap—the great lava flows, several thousand feet thick, that cover 250,000 square miles in India— built the lava dykes that cross South Africa, spread the Columbia Plateau in America, and laid the lava bed of the Pacific.

The Indian Ocean, from Java, an island full of volcanoes,

[1] Daly, *Our Mobile Earth*, p. 91.

extinct, dormant, and active, to Kilimanjaro, an extinct volcano over 19,000 feet high in East Africa, is circled with volcanoes and its bottom is of lava, with several volcanic isles in the middle of it. Along the Arabian coast of the Red Sea stretches a long chain of volcanoes; the numerous craters are all extinct, but it is not so long ago that they became inactive, the last eruptions having taken place in the year 1222 at Killis in northern Syria and in 1253 at Aden.[2]

In the Mediterranean region Thera (Santorin), which exploded with unusual force about 1500 B.C., is still active or dormant; Etna on Sicily, a snow-capped volcano, Stromboli, and Vulcano are active. On the mainland of Europe, however, the only active volcano left is Vesuvius. In the past France and the British Isles saw extensive volcanic activity, and though this activity is ascribed to the Tertiary, some of "the cones, craters, and lava-streams [in France] . . . stand out so fresh that they might almost be supposed to have been erupted only a few generations ago," in the words of Sir Archibald Geikie.[3]

Iceland in the North Atlantic has 107 volcanoes on it and thousands of craters, large and small; none of the volcanoes is geologically ancient, but many of them are extinct. The island is covered with coagulated lava, fissures, and crater formations. Iceland is one of the rare places where in modern times lava streams have been vomited from fissures in the earth without a crater having been formed.

From Iceland down the Atlantic, the Azores, the Canary Islands, the Cape Verde Islands, Ascension, and St. Helena are volcanic islands, some of them thrust up from the bottom of the ocean; their volcanic activity, like the activity of the many known volcanoes on the bottom of the Atlantic, has ceased.

In Patagonia volcanic eruptions have occurred down to fairly recent times, and the land between the Atlantic and the Andes is covered in many places with lava flows.

All in all only about four or five hundred volcanoes on earth are considered active or dormant, against a multiplicity of extinct cones. Yet only five or six hundred years ago many of the presently inactive volcanoes were still alive. This points to very

[2] Moritz, *Arabien, Studien zur physikalischen und historischen Geographie des Landes*, p. 12.
[3] A. Geikie, *The Ancient Volcanoes of Great Britain* (1897), p. viii.

great activity at a time only a few thousand years ago. At the rate of extinction witnessed by modern man, the greater part of the still active volcanoes will become inactive in a matter of several centuries.

The cause of volcanic activity is supposed to be in movements and fractures of the outer crust of the earth, "however these may be brought about, a matter as yet by no means settled." The coincidence in time and place of mountain folding and volcano building is regarded as significant for the solution of the problem of the origin of volcanoes.

Seas of lava and crater formations cover the entire face of the moon. "No one who has observed the moon, even through a relatively small telescope, can forget this picture of tremendous catastrophe: a flood of molten lava that has engulfed . . . and obliterated craters and mountain ridges in its path."[4] Whether the crater formations on the moon, some of which reach 150 miles in diameter, resulted from bombardment by enormous meteorites, or are extinct volanoes, or, as I assumed in *Worlds in Collision*, are the congealed effects of bubbling activity on the surface of the moon that became molten, the face of the moon is indisputable proof of catastrophic events on a planetary scale. The theory of uniformity can be taught only on moonless nights.

As with the volcanic activity, the seismic shocks, judged by their effects, must have been of a very different order of magnitude in the past. "The earthquakes of the present day," writes Eduard Suess in his *The Face of the Earth* (*Das Antlitz der Erde*), "are certainly but faint reminiscences of those telluric movements to which the structure of almost every mountain range bears witness. Numerous examples of great mountain chains suggest by their structure . . . episodal disturbances of such indescribable and overpowering violence, that the imagination refuses to follow the understanding. . . ."[5] Suess thought that mountain building came to an end before the advent of man; but today we know that it lasted well into Recent time, and consequently man must have witnessed the great earthquakes that made the globe shudder.

[4] O. Struve, review of *The Planets, Their Origin and Development*, by H. Urey, in *Scientific American*, August 1952.
[5] E. Suess, *The Face of the Earth* (1904), I, 17–18.

When the Andes rose in South America, according to the description of R. T. Chamberlin, "Hundreds, if not thousands, of cubic miles of the body of the earth almost instantaneously heaved upward produced a violent earthquake which spread . . . throughout the entire globe. Many world shaking earthquakes must have been by-products of the rise of the Sierras."[6]

Again, we now know that the Sierras attained their present height in the age of man, in Recent time.

And if we give credence to the records of earthquakes in the chronicles of the ancient East and in those of the classical age, we shall be amazed at the number of seismic shocks and tremors. One example is the Babylonian records on clay tablets stored in the library of Nineveh, excavated by Sir Henry Layard; another is the Roman records of a later age: in a single year during the Punic Wars (217 B.C.) fifty-seven earthquakes were reported in Rome.[7]

From all this it is apparent that seismic activity on our planet subsided very quickly in intensity as well as in the number of occurrences; and this again would point to a stress or stresses that took place not so long ago: earthquakes are readjustments of the terrestrial strata, with accompanying relief from the stress.

The theory of Alexis Perrey, regularly quoted in textbooks, connects the occurrence of earthquakes in our time with the position of the nearest celestial body, the moon. Earth tremors occur more often when the moon is full or when the earth is between the sun and the moon when the moon is new, or when it is between the sun and the earth; when the moon crosses the meridian of the afflicted locality; and when the moon is closest to the earth on its orbit. With the possible exception of the fourth case, the statistics for the last century appear to support Perrey's theory. But if this statistical theory is correct, then we have to look to the celestial sphere for the stresses that are relieved in earthquakes; and the farther in time from the stresses, the less numerous and less violent are the shocks.

Finally, a third natural phenomenon shows a definite downward curve. The number of comets visible to the unaided eye in recent centuries is only a small fraction of the number of

---

[6] Chamberlin in *The World and Man*, ed Moulton, p. 87.
[7] Pliny, *Natural History* (Trans. Bostock and Riley, 1855), II, 86.

cometary bodies that were observed in the historical past, in comparable periods of time. Whereas in our age about three comets are seen without the help of a telescope in the Northern Hemisphere in a century, in the days of imperial Rome, nineteen centuries ago, comets were so frequent that they were associated with many state events, such as the beginning of the rule of an emperor, his wars, his death. Often more than one comet was seen simultaneously. Some of the comets were spectacular and glowed even in the daytime.

Approaching the sun, a comet emits a tail consisting of gases and dust particles. It is believed that these tails are wasted and that their material does not return to the head. A comet that recurs every seventy-six years, as Halley's comet does, would have to grow and lose its tail about forty million times, if we take the usual figure for the age of the solar system, and such a wasting would have long ago reduced the comet to nothing.

In modern times, several comets of short period, or a period less than that of the Halley comet, and thus subject to check by observatories, vanished and did not return when expected; the number of comets, at least of those closely associated with the solar system, becomes even smaller.

According to a hypothesis offered by Swinne and referred to by H. Pettersson, "meteorites should be a relatively recent occurrence, limited to the last 25,000 years, and have been absent during preceding millions of years."[8]

The rapid decrease in luminosity of periodical comets points to some unusual activity in the sky in the geologically recent past; in the careful estimate of the Russian astronomer S. K. Vsehsviatsky (1953), this unusual activity took place in historical times, only a few thousand years ago.[9]

All three natural phenomena are on the wane. Volcanic activity is generally considered as connected with seismic activity; and the latter appears to be a response to a stress; and stress appears to have its origin in forces outside our earth.

[8] Pettersson, *Tellus* (*Quarterly Journal of Geophysics*), I (1949), 4.
[9] See reference to the work of S. K. Vsehsviatsky in the Supplement to this volume.

Chapter X

THIRTY-FIVE CENTURIES AGO

Clock Unwound

We can determine the time necessary for lakes to collect mud deposited by melting glaciers, for rivers to build their deltas, for waterfalls to cut their channels and to remove the bedrock, for lakes without outlets to accumulate their salts. We can ascertain how much time has passed since beaches were raised, by the state of their shells, and the age of volcanic rocks by the amount of erosion. By counting the annual bands of clay and silt we may find out the number of years spent in their deposition. By studying the rings in old tree trunks we can determine the time of climatic changes as reflected in their growth. The remains of extinct and extant animals—their appearance, position on the ladder of evolution, and state of fossilization—enable us to establish their time of existence. By the content of radiocarbon in organic matter we may detect the time when an animal or plant died, and by the accumulation of fluorine in bones the length of time since burial. Finally, by studying artifacts and archaeologically determinable strata in the lands of antiquity, we may discover the time of deposit of associated animal or human remains; and by associated pollens of plants, a geochronological scale of climatic changes can be formulated even for areas where no archaeologically datable objects are found.

There are a few other ways of calculating geological time: by measuring the amount of sediment on the bottom of the ocean; by computing the amount of salt in the oceans and comparing it with the annual influx of salts from land; and finally, by the analysis of rocks for their lead content as a product of decay of radioactive elements. But these ways, especially the last two, cannot be profitably applied for measuring time in thousands or tens of thousands of years; they were devised for reckoning time in millions of years.

Of the methods used to find how much time has passed since the ice cover started to melt, the "varve" method, until recently, was thought to be fairly precise. This method was introduced by G. de Geer, who counted the annual bands of silt and clay ("varves") deposited, coarse in summer and fine in winter, under the ice in the coastal lakes and rivers of Sweden, once covered by the glacial sheet of the Ice Age. De Geer calculated that it had taken about 5000 years to melt the ice cover from Schonen, at the southern tip of Sweden, to the place in the north where there are still glaciers in the mountains. In no place are there five thousand overlying varves; but De Geer looked for similar series or patterns of thick and thin varves from one lake to another, about fifteen hundred outcrops altogether, always with the thought that a varve series found high in the deposit of some southern lake would repeat itself closer to the bottom of a lake to the north.

Additional figures used in De Geer's evaluation of the time that passed since the end of the Ice Age are of a more hypothetical nature. For the preceding period, the time allegedly needed for the ice to retreat all the way, from Leipzig to southern Sweden, where no varves are found, De Geer offered, as a surmise, a span of 4000 years. Then he surmised further that the end of the melting of the ice cover coincided with the beginning of Neolithic time, which he placed 5000 years ago, thus arriving at the final figure of 14,000 years, or 12,000 years before the present era. The area of Stockholm was freed from ice about 10,000 years ago. Other scientists freely interpreted De Geer's data as indicating that the ice cover in Europe started to melt 25,000 or even 40,000 years ago.[1] The method, when applied to North America, also gave the figure the explorers looked for, namely 35,000 to 40,000 years; in this estimate great stretches

---

[1] Chamberlin, in *The World and Man*, ed. Moulton, p. 93; Daly, *Our Mobile Earth*, pp. 189–90; C. Schuchardt, *Vorgeschichte von Deutschland* (1943), p. 3.

of land without varves in them were freely evaluated as to the time in question.

De Geer applied his method of identifying synchronical varves to countries as far apart as Sweden, Central Asia, and South America. His telechronology was objected to on the ground that a dry phase in Scandinavia may not necessarily have coincided with a dry phase in the Himalayas or in the Andes, and that therefore the telechronology was built on an erroneous assumption.[2] But the method as applied to northern Europe or North America was hailed as a most exact geological time clock. The summing up of varves from one dried-out lake to another is a delicate process and often subjective appraisals replace an objective method; especially arbitrary are the estimates for intervening stretches of land where no varves are found.

In 1947 an ingenious new method of investigating the age of organic remains was developed by W. F. Libby of the University of Chicago. The radiocarbon dating method is based on the fact that when cosmic rays hit the upper atmosphere they break the nitrogen atoms into hydrogen (H) and radiocarbon (C_{14}), or carbon with two extra electrons, therefore unstable, or radioactive.

The radiocarbon is mixed with the atmospheric carbon and as carbon dioxide it is absorbed by plants; it enters the animal body that feeds on plants and also the carnivore that feeds on other animals. Thus all animal and plant cells as long as they live contain approximately the same amount of radiocarbon; when death comes, no new radiocarbon is assimilated and the radiocarbon present in the remains undergoes the process of decay, as every radioactive substance does. After 5568 years only half of the radiocarbon remains; after another 5568-year period only half of the half, or a quarter of the original content in the organic body, remains. A sample undergoing analysis—a piece of wood or skin—is burned to ashes and its radiocarbon content is determined by a Geiger counter. This method claims accuracy for organic objects between 1000 and 20,000 years old;

[2] E. Antevs, "Telecorrelation of Varve Curves," *Geologisma Förhandlingar*, 1935, p. 47; A. Wagner, *Klimaänderungen und Klimaschwankungen* (1940), p. 110.

bones and shells are unsuitable materials because organic carbon is easily lost in the process of fossilization, often being replaced by carbon in ground water and by mineral salts.

The first important result of the radiocarbon dating method in glacial chronology was a radical reduction of the terminal date of the Ice Age. It was shown that ice, instead of retreating 30,000 years ago, was still advancing 10,000 or 11,000 years ago.[3] This conflicts strongly with the figures arrived at by the varve method concerning the final phase of the Ice Age in North America.[4]

Even this great reduction of the date of the end of the Ice Age is not final. Radiocarbon analysis, according to Professor Frederick Johnson, chairman of the committee for selection of samples for analysis,[5] revealed "puzzling exceptions." In numerous cases the shortening of the time schedule was so great that, as the only recourse, Libby assumed a "contamination" by radiocarbon. But in many other cases "the reason for the discrepancies cannot be explained." Altogether the method indicates that "geological developments were speedier than formerly supposed."[6]

H. E. Suess of the United States Geological Survey reported recently that wood found at the base of interbedded blue till, peat, and outwash of drift, and ascribed by its finder to the Late Wisconsin (last) glaciation, is, according to radiocarbon analysis, but 3300 years old (with a margin of error up to two hundred years both ways), or of the middle of the second millennium before the present era. Still more recently Suess and Rubin reported that "a glacial advance in the mountains of western United States was determined to have occurred about 3000 years ago."[7]

Already there is an accumulation of similar results that do not fit into the accepted scheme, even if the Ice Age is brought as close to our time as 10,000 years. Professor Johnson says:

[3] F. Johnson in Libby, *Radiocarbon Dating* (1952), p. 105.

[4] Antevs, "Geochronology of the Deglacial and Neothermal Ages," *Journal of Geology*, LXI (1953), 195–230. Cf., however, G. de Geer in *Geografiska Annaler*, 1926. H. 4. He evaluated the time when the ice cover left the region of Toronto as about 9750 years ago.

[5] The Committee on Carbon 14 of the American Anthropological Association and the Geological Society of America.

[6] Johnson in Libby, *Radiocarbon Dating*, pp. 97, 99, 105.

[7] *Science*, September 24, 1954, and April 8, 1955.

"There is no way at the moment to prove whether the valid dates, the 'invalid ones,' or the 'present ideas' are in error."[8] He says also: "Until the number of measurements can be increased to a point permitting some explanation of contradictions with other apparently trustworthy data, it is necessary to continue to form judgments concerning validity by a combination of all available information."

With this idea in mind, I offer in the following sections a review of the results of several other methods of time measurement, especially as regards the dating of the last glaciation.

Libby recognizes that the exactness of his method is dependent on two assumptions. The first is that for the last 20,000 or 30,000 years the amount of cosmic radiation reaching our atmosphere remained constant; the other is that the quantity of water in the oceans has not changed in the same period of time. Actually only a minor part of the radiocarbon created by cosmic rays is absorbed by plant and animals, the so-called biosphere; a still smaller part is present in the atmopshere; the largest share is absorbed by the ocean.

Libby stressed the significance of these factors. It transpires that if there were cosmic catastrophes in the past cosmic radiation could have reached the earth at a different intensity; and in a future book I intend to show that the waters of the oceans and their salts were increased substantially in a recent geological age.

Bearing in mind these limitations, I confidently expect that in the field of geology more and more "puzzling" results of radiocarbon tests will compel a full-scale revision of the dating of the glacial periods.[9]

The Glacial Lake Agassiz

Lake Agassiz, the largest glacial lake of North America, once covered the region at present occupied by Lake Winnipeg, Lake Manitoba, a number of other lakes in Canada, and parts of the North Central States of the United States. It exceeded

---

[8] Johnson in Libby, *Radiocarbon Dating*, p. 106.

[9] In the field of archaeology, I expect the radiocarbon tests to confirm that the time of the Eighteenth Dynasty in Egypt must be reduced by five to six hundred years, and the time of the Nineteenth and Twentieth Dynasties by a full seven hundred years, as I maintain in *Ages in Chaos*.

the aggregate area of the five Great Lakes tributary to the St. Lawrence River. It was formed when the ice of North America melted. Study of its sediment, however, disclosed that its entire duration had been definitely less than one thousand years, a measure of time unexpectedly short; this indicates also that the glacial cover melted under catastrophic conditions. Warren Upham, the American glaciologist, wrote: "The geologic suddenness of the final melting of the ice-sheet, proved by the brevity of existence of its attendant glacial lakes, presents scarcely less difficulty for explanation of its causes and climatic conditions than the earlier changes from mild and warm pre-glacial conditions to prolonged cold and ice accumulation."[1]

Not only was the life of the glacial Lake Agassiz measured in hundreds of years and the melting of the continental ice cover that gave rise to this lake, of short duration but this melting must have taken place only recently: the erosion on the shores of Lake Agassiz indicates that it existed only a short time ago. Upham also recognized that the shoreline of the extinct lake is not horizontal, which indicates that the warping too occurred recently.

Although this study of Lake Agassiz by Upham is over fifty years old, its conclusions have never been challenged. He also stated:

"Another indication that the final melting of the ice sheet upon British America was separated by only a very short interval, geologically speaking, from the present time is seen in the wonderfully perfect preservation of the glacial striation and polishing on the surface of the more enduring rocks. . . . It seems impossible that these rock exposures can have so well withstood weathering in the severe climate of those northern regions longer than a few thousand years at the most."[2]

Upham realized and stressed that "these measures of time" are "surprisingly short, whether we compare them on the one hand with the period of authentic human history or on the other hand with the long record of geology."

How it started, how it ended—all appears enigmatic; what is clear is that great changes took place but a few thousand years ago, under catastrophic conditions.

[1] Warren Upham, *The Glacial Lake Agassiz* (1895), p. 240.
[2] Ibid., p. 239

When Lyell, on his trip to the United States, visited Niagara Falls, he talked with someone who lived in the vicinity and was told that the falls retreat about three feet a year. Since the natives of a country are likely to exaggerate, Lyell announced that one foot per annum would be a better figure. From this he concluded that over thirty-five thousand years were necessary, from the time the land was freed from the ice cover and the falls started their work of erosion, to cut the gorge from Queenston to the place it occupied in the year of Lyell's visit. Since then this figure has often been mentioned in textbooks as the length of time from the end of the glacial period.

The date of the end of the Ice Age was not changed when subsequent examination of records indicated that since 1764 the falls had retreated from Lake Ontario toward Lake Erie at the rate of five feet per year, and that, if the process of wearing down the rock had gone on at the same rate from the time of the retreat of the ice cover, seven thousand years would have been sufficient to do the work. However, since in the beginning, when the ice melted and a swollen stream carried the detritus abrading the rock of the gorge, the erosion rate must have been much more rapid, the age of the gorge must be further reduced. According to G. F. Wright, author of *The Ice Age in North America*, five thousand years may be regarded as an adequate figure.[1] The erosion and sedimentation of the shores of Lake Michigan also suggest a lapse of time reckoned in thousands, but not tens of thousands, of years since the beginning of the process.[2]

In the 1920s, however, when borings were made for a railroad bridge, it was found that the middle part of Whirlpool Rapids Gorge of Niagara Falls contained a thick deposit of glacial boulder clay, indicating that it had been excavated once, had been filled with drift, and then partly re-excavated by the falls in post-glacial times.[3] While the question of the age of the

[1] G. F. Wright, "The Date of the Glacial Period," *The Ice Age in North America and Its Bearing upon the Antiquity of Man*.

[2] E. Andrews, *Transactions of the Chicago Academy of Sciences*, Vol. II.

[3] W. A. Johnston, "The Age of the Upper Great Gorge of Niagara River," *Transactions of the Royal Society of Canada*, Ser. 3, Vol. 22, Sec. 4, pp. 13–29; F. B. Taylor, *New Facts on the Niagara Gorge*, Michigan Academy of Sciences, XII (1929), 251–65.

falls thus becomes complicated, the discovery shows that the post-glacial period was of much shorter duration than generally assumed, even if the rate of retreat of the falls is reduced to the minimum figure of under four feet per year, as observed in more recent years. R. F. Flint of Yale writes:

"We are obliged to fall back on the Upper Great Gorge, the uppermost segment of the whole gorge, which appears to be genuinely postglacial. Redeterminations by W. H. Boyd showed the present rate of recession of the Horseshoe Falls to be, not five feet, but rather 3.8 feet, per year. Hence the age of the Upper Great Gorge is calculated as somewhat more than four thousand years—and to obtain even this [low] figure we have to assume that the rate of recession has been constant, although we know that discharge has in fact varied greatly during postglacial times."[4] If due allowance is made for this last factor, the age of the Upper Great Gorge of Niagara Falls would be somewhere between 2500 and 3500 years. It follows that the ice retreated in historical times, somewhere between the years 1500 and 500 before the present era.

The Rhone Glacier

The lifetime of a glacier is determined by measuring the detritus deposited by the melting ice. Albert Heim, the Swiss naturalist, estimated the age of the glacial river Muota that flows into Lake Lucerne at sixteen thousand years. F. A. Forel, another Swiss naturalist, undertook an evaluation of the detrital mud deposited by the Rhone Glacier on the bottom of Lake Geneva. He arrived at a figure close to twelve thousand years as the span of time necessary for the mud and detritus to have been deposited on the bottom of the lake, or from the height of the Ice Age to the present. Forel's result actually signifies that the Rhone Glacier, which feeds the river and the lake, is evidence of the short duration of the postglacial period, or even of the entire Ice Age if the origin of the lake goes back to the first

[4] Flint, *Glacial Geology and the Pleistocene Epoch*, p. 382. C. W. Wolfe, professor of geology at Boston University, in *This Earth of Ours, Past and Present* (1949), writes (p. 176): "A rather satisfactory estimate on the recession of the Horseshoe Falls section indicates that the falls are moving upstream at the surprising rate of five feet per year. . . ."

glacial period. These estimates, when announced, were much lower than expected.

The eminent French geologist of the beginning of this century, and a colleague of Heim and Forel, A. Cochon de Lapparent, arrived at an even more radical result. In the time of its greatest expansion, the Rhone Glacier reached from Valais to Lyons. De Lapparent took the average figure of progression as seen today on larger glaciers. Mer de Glace, a glacier on Mont Blanc, moves fifty centimetres in twenty-four hours. Moving at a comparative velocity, the Rhone Glacier, when expanding, would have required 2475 years to progress from Valais to Lyons. Then, comparing the terminal moraines, or stone and detritus accumulation, of several present-day glaciers with the moraines left by the Rhone Glacier at its maximum expansion, De Lapparent again arrived at a figure of about 2400 years. He also concluded that the entire Ice Age was of very short duration. To this another geologist, Albrecht Penck, objected.[1] His objection was based not on a disproval of the above figures, but on a claim that great evolutionary changes took place during the consecutive interglacial periods. The divergence of opinion between them was so great that hundreds of thousands of years in Penck's scheme were reduced to mere thousands of years in De Lapparent's calculations. Penck estimated the duration of the Ice Age, with its four glacial and three interglacial periods, as one million years. Each of the four glaciations and deglaciations must have consumed one hundred thousand years and more. The argument for his estimate is this: How much time was necessary to produce the changes in nature, if no catastrophes intervened? And how long would it take to produce changes in animals by means of a process that in our own day is so slow as to be almost imperceptible?

Carl Schuchardt, in his book, *Alteuropa*, warned his colleagues not to turn deaf ears to voices like that of De Lapparent. Let us assume that the geological processes were always as they are now. In Ehringsdorf near Weimar there is a tufa layer in which, during the entire last interglacial period, calcium was deposited. "But should we even assume all kinds of imaginable causes that would have retarded the deposition of calcium so as to make it

[1] A. Penck, "Das Alter des Menschengeschlechts," *Zeitschrift für Ethnologie*, XL (1908), 390ff.

ten times as slow as at present, still we would have only 3000 years and not 100,000!"[2]

If we follow the principle of quantitative analysis and accept De Lapparent's figure as approximately correct, the maximal extension of the Rhone Glacier dates from a point well within the bounds of human history.

The recent field work in the Alps actually revealed that numerous glaciers there are no older than 4000 years. This startling discovery made the following statement necessary: "A large number of the present glaciers in the Alps are not survivors of the last glacial maximum, as was formerly universally believed, but are glaciers newly created within roughly the last 4000 years."[3]

The Mississippi

The Mississippi carries yearly in its stream many billions of tons of detritus, a large part of which is deposited in the delta. As early as 1861, Humphreys and Abbot calculated the age of the Mississippi by evaluating the detritus borne by it and the sediment deposited in the delta. They arrived at the low figure of 5000 years as the age of the delta, its birth thus being related to about the year 2800 before the present era.[1] However, when at the close of the Ice Age the ice cover melted in the north, multitudinous streams must have carried an enormous amount of detritus into the Mississippi and its tributary, the Missouri, and for this reason the above figure, if otherwise properly calculated, must be appreciably reduced. It is assumed that when the continental ice started to melt and the Great Lakes became swollen, but the St. Lawrence was still blocked by ice, the water of the basin emptied to a great extent into the Gulf of Mexico through the Mississippi.

The Falls of St. Anthony on this stream at Minneapolis have excavated a long gorge by removing the bedrock. In the 1870s and 1880s N. H. Winchell made these falls the subject of a

[2] *Alteuropa* (1929), p. 16; Idem, *Vorgeschichte von Deutschland* (1943), p. 3.
[3] Flint, *Glacial Geology*, p. 491. Cf. R. von Klebelsberg, *Geologie von Tirol* (1935), p. 573.
[1] Humphreys and Abbot, *Report on the Mississippi River* (1861), a publication of the U.S. Army.

study. Comparing topographical maps covering two hundred years, he concluded that the falls had retreated 2.44 feet yearly. If this was the constant rate of retreat, the falls must have started 8000 years ago.[2] But here, too, a larger stream carrying abundant detritus, which abraded the bedrock, must have flowed when the ice cover melted. J. D. Dana, studying the area of Lake Champlain and of the Northeastern states in general, came to the conclusion that prodigious floods of almost unimaginable magnitude accompanied the melting of the ice cover: in the lower part of the Connecticut River the floods rose two hundred feet above the present high-water mark.[3] And if this is true for those regions, it must be true also for the valley of the Mississippi. Consequently the gorge of the Falls of St. Anthony must be of more recent date than Winchell calculated, though even his figure was regarded as much too low.

The protracted discussion of the results derived from the exploration of Niagara and St. Anthony falls demonstrated the need for yet another area of investigation, preferably the delta of a stream fed by a still existing glacier that could be carefully studied. For that purpose the delta of the Bear River was selected (a stream from a melting glacier that enters the Portland Canal on the Alaska-British Columbia border). On the basis of three earlier accurate surveys made between the years 1909 and 1927, G. Hanson in 1934 calculated with great exactness the annual growth of the delta through deposited sediment. At the present rate of sedimentation the delta is estimated to be "only 3600 years old."[4] The glacier that feeds the Bear River was formed and began to melt in the middle of the second millennium before the present era.

Fossils in Florida

On the Atlantic coast of Florida, at Vero in the Indian River region, in 1915 and 1916, human remains were found in associa-

[2] *Minnesota Geologic and Natural History Survey for 1876* (1877), pp. 175–89.
[3] G. F. Wright, *The Ice Age in North America*, p. 635.
[4] G. Hanson, "The Bear River delta, British Columbia, and its significance regarding Pleistocene and Recent glaciation," *Royal Society of Canada, Transactions*, Ser. 3, Vol. 28, Sec. 4, pp. 179–85. See also Flint, *Glacial Geology*, p. 495.

tion with the bones of Ice Age (Pleistocene) animals, many of which either became extinct, like the sabre-toothed tiger, or have disappeared from the Americas, like the camel.

The find caused immediate excitement among geologists and anthropologists. Beside the human bones pottery was found, as well as bone implements and worked stone. Aleš Hrdlička, of the Smithsonian Institution of Washington, D.C., a renowned anthropologist (who generally opposed the view that man existed in America in the Ice Age), wrote that the "advanced state of culture, such as that shown by the pottery, bone implements, and worked stone brought from a considerable distance, implies a numerous population spread over large areas, acquainted thoroughly with fire, with cooking food, and with all the usual primitive arts"; the human remains and relics could not be of an antiquity "comparable with that of fossil remains with which they are associated."[1] He also published the opinion of W. H. Holmes, head curator of the Department of Anthropology of the United States National Museum, who investigated the pottery obtained by Hrdlička from Vero. These were bowls "such as were in common use among the Indian tribes of Florida." When compared with vessels from Florida earth mounds, "no significant distinction can be made; in material, thickness of walls, finish of rim, surface finish, colour, state of preservation, and size and shape," the vessels "are identical." There thus appears "not the least ground in the evidence of the specimens themselves for the assumption that the Vero pottery pertains to any other people than the mound-building Indian tribes of Florida of the pre-Columbian time."

But the bones of man and his artifacts (pottery) were found among the extinct animals. The discoverer of the Vero deposits, E. H. Sellards, state geologist of Florida and a very capable paleontologist, wrote in the debate that ensued: "That the human bones are fossils normal to this stratum and contemporaneous with the associated vertebrates is determined by their place in the formation, their manner of occurrence, their intimate relation to the bones of other animals, and the degree of mineralization of the bones." This "degree of mineralization of the human bones is identical with that of the associated bones

[1] "Preliminary Report on Finds of Supposedly Ancient Human Remains at Vero, Florida," *Journal of Geology*, XXV (1917).

of the other animals." In his view the evidence obtained "affords proof that man reached America at an early date and was present on the continent in association with a Pleistocene [Ice Age] fauna."[2] Anthropologists of the Hrdlička school would not accept this, claiming a late arrival of man on the American continent, and the presence of pottery was in their view proof of a late date for the human bones. The human skulls, though fossilized, did not differ from the skulls of the Indians of today.

In 1923–29, thirty-three miles north of Vero, in Melbourne, Florida, another such association of human remains and extinct animals was found, "a remarkably rich assemblage of animal bones, many of which represent species which became extinct at or after the close of the Pleistocene [Ice Age] epoch."[3] The discoverer, J. W. Gidley, of the United States National Museum, established unequivocally that in Melbourne—as in Vero—the human bones were of the same stratum and in the same state of fossilization as the bones of the extinct animals. And again human artifacts were found with the bones. The "projectile points, awls, and pins" found with the human bones at Melbourne as well as at Vero are of the same workmanship as those unearthed in early Indian sites, two thousand of which are known in the area.

All these and other considerations of an anthropological as well as geological nature, being summed up, prove, in the opinion of I. Rouse, a recent analyst of the much-debated fossils of Florida, that "the Vero and Melbourne man should have been in existence between 2000 B.C. and the year zero A.D."[4] This does not solve the problem of the association of extinct animals and man who lived between two and four thousand years ago, in the second and first millennia before the present era.

There is no proper way out of this dilemma, other than the assumption that now extinct animals still existed in historical

[2] "On the Association of Human Remains and Extinct Vertebrates at Vero, Florida," *Journal of Geology*, XXV (1917).
[3] J. W. Gidley, "Ancient man in Florida," *Bulletin of the Geological Society of America*, Vol. XL, pp. 491–502; J. W. Gidley and F. B. Loomis, "Fossil man in Florida," *American Journal of Science*, 5th Ser., Vol. 12, pp. 254–65.
[4] I. Rouse, "Vero and Melbourne Man," *Transactions of the New York Academy of Sciences*, Ser. II, Vol. 12 (1950), pp. 224ff.

times and that the catastrophe which overwhelmed man and animals and annihilated numerous species occurred in the second or first millennium before the present era.

The geologists are right: the human remains and artifacts of Vero and Melbourne in Florida are of the same age as the fossils of the extinct animals.

The anthropologists are equally right: the human remains and artifacts are of the second or first millennium before the present era.

What follows? It follows that the extinct animals belonged to the recent past. It follows also that some paroxysm of nature heaped together these assemblages; the same paroxysm of nature may have destroyed numerous species so that they became extinct.

Lakes of the Great Basin and the End of the Ice Age

The Sierra Nevada chain rises between the Great Basin to the east and the Pacific, cutting off the drainage to the ocean. Abert and Summer lakes in southern Oregon have no outlets. They are regarded as remnants of a once large glacial lake, Chewaucan. W. van Winkle of the United States Geological Survey investigated the saline content of these two lakes and wrote: "A conservative estimate of the age of Summer and Abert Lakes, based on their concentration and area, the composition of the influent waters, and the rate of evaporation, is 4000 years."[1] If this conclusion is correct, the post-glacial epoch is no longer than 4000 years. Startled at his own result, van Winkle conjectured that salt deposits of the early Chewaucan Lake may be hidden beneath the bottom sediments of the present Abert and Summer lakes.

To the east of Sequoia National Park and Mount Whitney in California lies Owens Lake. It is supplied by the Owens River and it has no outlet. At some time in the past the surface level of the lake, because of a greater water supply, was so much higher that it overflowed its basin. H. S. Gale analyzed the water of the lake and of the river for chlorine and sodium and

[1] Walton van Winkle, "Quality of the Surface Waters of Oregon," U.S. Geological Survey, Water Supply Paper 363 (Washington, 1914).

came to the conclusion that the river required 4200 years to supply the chlorine present in the lake and 3500 years to supply its sodium. Ellsworth Huntington of Yale found these figures too high, because no allowance was made for greater rainfall and "freshening of the lake" in the past, and consequently he reduced the age of the lake to 2500 years, which would place its origin not far from the middle of the first millennium before the present era.[2]

Another vast lake of the past without an outlet to the sea was Lake Lahontan in the Great Basin of Nevada, which covered an area of 8500 square miles. As its water level fell, it split up into a number of lakes divided by a desert terrain. In the 1880s I. Russell of the United States Geological Survey investigated Lake Lahontan and its basin, and established that the lake was never completely dried out and that the present-day Pyramid and Winnemucca lakes north of Reno and Walker Lake southwest of it are the residuals of the older and larger lake.[3] He concluded that Lake Lahontan existed during the Ice Age and was contemporaneous with the different stages of glaciation of that age. He also found bones of Ice Age animals in the deposits of the ancient lake.

More recently, Lahontan and its residual lakes were explored anew by J. Claude Jones, and the results of his work were published as "Geological History of Lake Lahontan" by the Carnegie Institution of Washington.[4] He investigated the saline content of Pyramid and Winnemucca lakes and of the Truckee River that feed them. He found that the river could have supplied the entire content of chlorine of these two lakes in 3881 years. "A similar calculation, using sodium instead of chlorine, gave 2447 years necessary." Jones's careful work led him to agree with Russell that Lake Lahontan never fully dried up and that the existing lakes are its residuals.

But these conclusions require that the age of the mammals of the Ice Age, found in the deposits of Lake Lahontan, be not greater than that of the lake. This means that the Ice Age ended only twenty-five to thirty-nine centuries ago. Jones checked the

[2] *Quaternary Climates*, monographs by J. Claude Jones, Ernst Antevs, and Ellsworth Huntington (Carnegie Institution of Washington, 1925), p. 200.
[3] I. Russell, "Geologic History of Lake Lahontan," U.S. Geological Survey, Monograph 11 (1886).
[4] Jones, Antevs, and Huntington, *Quaternary Climates*.

figures from the rate of accumulation of chlorine and sodium as brought in by the Truckee River, with other methods, such as the accumulation of chlorine in lakes during the thirty-one years that had passed since the analysis made by Russell, and also the rate of concentration of salts by evaporation, and each time reached the result that the entire history of Pyramid and Winnemucca lakes "is within the last 3000 years."[5]

Bones of horses, elephants and camels, animals that became extinct in the Americas, were found in the Lahontan sediments, as well as a spear point of human manufacture.[6] When a branch of the Southern Pacific Railroad was laid through Astor Pass, a large gravel pit of Lahontan age was opened, and J. C. Merriam of the University of California identified among the bones the skeletal remains of *Felix atrox*, a species of lion found also in the asphalt pit of Rancho la Brea, as well as a species of horse and a camel, also found in La Brea.[7] "All of these forms are now extinct and neither camels nor lions are found on this continent as a part of the present native fauna."[8] The similarity of the fauna of the asphalt pits of La Brea and the deposits of Lake Lahontan led Merriam to decide that they were contemporaneous.

On the basis of his analyses Jones came to the conclusion that the extinct animals lived in North America into historical times. This was an unusual statement and it was opposed at first on the ground that his interpretation of his observations was "obviously erroneous, since [it] led him to the conclusion that the mastodon and the camel lived on in North America into historical times."[9] But this is an argument of a preconceived nature, not based on findings of field geology. Either the Ice Age animals survived the Ice Age, or some of the vicissitudes of the Ice Age occurred in historical times.

[5] Jones, in *Quaternary Climates*, p. 4.
[6] Russell, U.S. Geological Survey, Monograph 11, p. 143.
[7] J. C. Merriam, *California University Bulletin*, Department of Geology, VIII (1915), 377–384.
[8] Jones, in *Quaternary Climates*, pp. 49–50.
[9] Brooks, *Climate through the Ages* (2nd ed.; 1949), p. 346.

Chapter XI

KLIMASTURZ

Klimasturz

Not long ago "it was generally believed that variations of climate came to an end with the Quaternary Ice Age, a period, moreover, which was placed hundreds of thousands years ago."[1] It was regarded as an established fact in the history of climate and in historical geology that during the period since the close of the glacial ages, called Recent, the climate of the earth did not change appreciably.

Then, in 1910, at the International Geological Congress in Stockholm, facts were placed before the scientists that demonstrated great changes and catastrophic fluctuations in the climate of the earth in the past few thousand years. Since that congress many works have been written to describe the climatic as well as geological changes that occurred in this recent time. In many places the present land was covered by sea and the present sea was land. For instance, from the changes in the mollusc population of the seas and the tree population of the submerged forests, it was concluded that the North and the Baltic seas assumed their present shapes during the Recent period. Explorations conducted in various countries combined also to demonstrate that "the ice age itself was not so remote as it had seemed to be, and that in fact the post-glacial 'geology' of Europe was partly contemporaneous with the 'history' of Egypt."[2]

One very strong disturbance in the climate, or climatic plunge (*Klimasturz*,) occurred in the Subboreal, a subdivision

[1] Brooks, *Climate through the Ages* (2nd ed.), p. 281.
[2] Ibid.

of the Recent, and is assigned to the middle of the second millennium before the present era. The second climatic catastrophe of the Recent period took place in the century following the year 800 B.C., a time period that is well within recorded history. "The beginning of the 'period of unchanging climate' has advanced later and later before the attacks of geologists, and now, in the minds of most of the authors who concern themselves with the subject, it apparently stands only a few centuries before Christ."[3]

The new understanding originated with Axel Blytt, a Norwegian scientist who began his work in the seventies of the last century. Gunnar Andersson and Rutger Sernander, also Scandinavian scientists, carried on the work that Blytt started. Thus it happened that Scandinavia and the surrounding seas were investigated first.

In Scandinavia the last *Klimasturz* marked the end of the Bronze Age. The following centuries offer a picture of desolation and wretchedness attributed to the altered climate. "Opulent plenty" was followed by "striking poverty."[4] Study of changes in the flora, as reflected in the pollens of trees found in the ancient moors, also disclosed a picture of a sudden climatic catastrophe. "The deterioration of the climate must have been catastrophic in character," wrote Sernander, whose laboratory at the University of Uppsala became the centre of research in the history of climate. To the period of the greatest change he gave the name Fimbul-Winter, borrowing the term from the northern epic, the *Edda*. In this epic Fimbul-Winter is a designation for a snowfall that continued through winter and summer alike, uninterrupted for years.

The last series of climatic disturbances of the eighth and the beginning of the seventh centuries did not take the form of a single drop in temperature. According to Sernander, "The desolating effect of the Fimbul-Winter on the northern culture was caused not so much by the fall in temperature as by oscillations and instability of the climate. . . ."[5] However, its catastrophic beginning was emphasized by him and also by other authors; thus G. Kossinna, who ascribes the *Klimasturz* to

[3] Ibid.
[4] R. Sernander, "Klimaverschlechterung, Postglaciale" in *Reallexikon der Vorgeschichte*, ed. Max Ebert, VII (1926).
[5] Ibid.

"about the year 700 B.C.," stressed that it took place with catastrophic suddenness.[6]

Tree Rings

The annual rings of trees reveal whether in some particular year or period growth was stimulated or inhibited. The oldest trees on record are among the sequoias of California. Some of them measure ninety feet in circumference. Of all the specimens whose rings were counted, the most ancient started life after the year 1300 before the present era. (The age of the General Sherman tree in Sequoia National Park is not known, since it has not been cut down.) Thus it appears that no tree has survived to modern times from the days of the great catastrophe of the middle of the second millennium. The sequoias are protected against fire by a bark often two feet thick, which resists combustion almost as well as asbestos. In order to survive through the days of global catastrophe a tree had also to withstand hurricane and tidal wave, and live in a sunless world under a canopy of dust clouds that enshrouded the world for many years.

The oldest trees that started life about 3200 years ago offer insight into the influence on their growth, as caused by a series of later climatic disturbances on the global scale that, according to the pollen analysis, took place in the eighth and the beginning of the seventh centuries, or 2700 years ago. According to the historical material collected in *Worlds in Collision*, the memorable dates are 747, 702, and especially 687 B.C.

The Carnegie Institution published in 1919 a graph drawn by A. E. Douglass, then director of Steward Observatory, who studied tree rings in order to discover the solar activity of the past.[1] The graph actually reveals a spurt of oscillations in the annual growth of the tree rings around the year 747 B.C. (the identification of the rings as to their years is approximate). There is an unusually high crest in the last years of the eighth century and the beginning of the seventh century. After a record high crest of six-year duration there is in 687 B.C. a precipitate drop.

[6] G. Kossinna in *Mannus, Zeitschrift für Vorgeschichte*, IV (1912), 418.
..[1] A. E. Douglass, *Climatic Cycles and Tree Growth*, Carnegie Institution Publications, No. 289 (1919), L, 1118–19.

Natural upheavals of great violence reacted destructively upon the forests. But those trees that survived the *Klimastürze* of the eighth and seventh centuries (hurricanes, floods, lava, and fire) were stimulated to growth by the increased presence of carbon dioxide in the air, yet impeded by a screen of clouds and dust; they might have been invigorated by electrical discharges in the atmosphere and possibly magnetic storms, and benefited from the addition of ashes to the soil. The singeing of leaves and changed conditions of ground water, as well as the change of climate generally, must have entered the picture. All in all, strong oscillations in the size of tree rings must be expected in years of great natural catastrophes. These are clearly recognizable on the annual rings of sequoias formed about the years 747, 702, 687 B.C., and generally in that century.

Lake Dwellings

At the close of the Stone Age in Europe, about 1800 B.C., lake dwellings existed in which man and his cattle lived, protected from wild animals. The structures were erected on wooden poles driven into the ground. Remains of such dwellings were discovered on the shores of the lakes of Scandinavia, Germany, Switzerland, and northern Italy. Sometime in the middle of the second millennium before the present era a "high-water catastrophe" occurred. The villages were overwhelmed and covered with mud, sand, and calcareous deposit. Life came to an end in all lake dwellings. Then for about three or four centuries they were not rebuilt; but after 1200 B.C. new villages were erected, in some places on top of the earlier ones, in other places on new ground. It was already the Bronze Age in Europe; bronze articles are found among the remains of the lake dwellings of that period.

After a second period of prosperity, which lasted for about four centuries, in the eighth century before the present era a new catastrophe overwhelmed the lake villages on all the lakes of central and northern Europe, and again it was a "high-water catastrophe"; once more mud and sand covered the villages on poles, and, abandoned by man, they were never rebuilt.

Thus it occurred that twice, once at the close of the Stone (Neolithic) Age and the second time at the close of the Bronze

Age, the lake dwellings were swamped by water and mired in mud. The coincidence of their destruction with the end of the cultural ages was called *merkwürdig* (remarkable) by Ischer, who explored the Bielersee (Lake of Bienne),[1] and *rätselhaft* (puzzling) by Reinerth, who explored the Bodensee (Lake Constance);[2] but all explorers agree that the cause was a natural catastrophe at the end of the Stone Age and another natural catastrophe before the advent of the Iron Age in central and northern Europe. It is also generally held that the catastrophes were accompanied by very great and sudden climatic changes, *Klimastürze*.[3] For the first event scientists fix the date at about 1500 B.C., some diverging by a few centuries either way, from 1800 B.C., to 1400 B.C.[4] For the second event the preferred date is the eighth century before the present era,[5] with some authors reducing the date to the seventh century.

H. Gams and R. Nordhagen made an extensive survey of German and Swiss lakes and fens and published a classical work on the subject.[6] They found not only that the lakes at two periods in the past—the end of the Neolithic (recent Stone) Age in Europe, in the middle of the second millennium, and in the eighth century before the present era—were subjected to high-water catastrophes, but also that these catastrophes were accompanied or caused by very strong tectonic movements. The lakes suddenly lost their horizontal position, one end often being tilted up, the other down, so that the old strand line may now be seen to run obliquely to the horizon. Such is the case of Ammersee and Würmsee in the foothills of the Bavarian Alps and of other lakes on the Alpine fringes.[7] In these catastrophes, the water of the Bodensee (Lake Constance) rose thirty feet, and the bed was tilted. The tilted strand lines of lakes were also

[1] T. Ischer, *Die Pfahlbauten des Bielersees*, p. 99.

[2] H. Reinerth, *Die Pfahlbauten am Bodensee* (1922), p. 35.

[3] O. Paret, *Das Neue Bild der Vorgeschichte* (1948), p. 44.

[4] Brooks, *Climate through the Ages* (2nd ed.), p. 300.

[5] Paret, *Das Neue Bild der Vorgeschichte*, p. 135. In the first edition of his book, *Climate through the Ages*, Brooks placed the beginning of the Sub-atlantic time, that followed the last *Klimasturz*, in 850 B.C. and in the second edition in the end of the sixth century before the present era.

[6] H. Gams and R. Nordhagen, "Postglaziale Klimaänderungen und Erdkrustbewegungen in Mitteleuropa," *Mitteilungen der Geographischen Gesellschaft in München*, XVI, Heft 2 (1923), 13-348. [7] Ibid., pp. 17-44.

found in regions far away from the Alps, for instance, in Norway by Bravais and Hannsen and in Sweden by De Geer and Sandegren, dating from the same ages.[8]

Some lake basins were suddenly emptied of all their water as the result of the tilting, as were Ess-see and Federsee.[9] The Isartal (the valley of the Isar) in the Bavarian Alps was "violently torn out" in "very recent times."[10] And in the Inntal in the Tyrol the "many changes of river beds are indicative of ground movements on a great scale."[11]

All the explored lakes of the Swiss Alps region, as well as of the Tyrol, the Bavarian Alps, and around the Jura, were flooded twice in catastrophic surges of water (*Hochwasserkatastrophen*), and the cause lay in tectonic movements and in the sudden melting of glaciers. It happened in the post-glacial period, the last time actually in the historical age, not long before the Romans started to spread into those parts of the world.[12]

Gams and Nordhagen also presented extensive material to show that the tectonic disturbances were accompanied not only by high-water catastrophes but also by climatic changes. They undertook a close examination of the pollen content of peat bogs. Since the pollen of each species of tree is characteristic, it is possible to detect by analysis what kinds of forests grew in various periods of the past, and consequently the then prevailing climate. The pollen disclosed a "radical change of life conditions, not a slow building of fens."[13] Men and animals suddenly disappeared from the scene, although at that time the area was already rather thickly populated. Oak was replaced by fir, and fir descended from the heights on which it had grown, leaving them barren.

The Alpine passes were much travelled during the Bronze Age: many bronze objects from before 700 B.C. were found in numerous places, especially on St. Bernard. Also mines were worked in the Alps in the Bronze Age. With the advent of the *Klimasturz* the mines were suddenly abandoned, and the passes were not travelled any longer, as though life in the Alps had been extinguished.[14]

[8] Ibid., pp. 34, 225–42.   [9] Ibid., p. 44.   [10] Ibid., pp. 53, 60.
[11] Ibid., p. 73.   [12] Ibid., p. 219.   [13] Ibid., p. 94.
[14] Cf. the section "Der vorgeschichtliche Verkehr über die Alpenpässe" in the quoted work by Gams and Nordhagen.

A chronological scale has been set up relating pollen analysis to archaeological finds. The pollen analysis, like other methods of investigation, showed that in the middle of the second millennium and again in the eighth or seventh century before the present era central Europe and Scandinavia passed through climatic catastrophes.

Coincident tectonic, high-water, and climatic catastrophes thus brought havoc to the entire area investigated, from Norway to the Jura, the Alps, and the Tyrol, tearing out valleys, overturning lakes, annihilating human and animal life, suddenly changing the climate, replacing forests with bogs, and doing this at least twice in Subboreal time, the period that is estimated to have lasted from about the year 2000 B.C. or possibly from a date closer to the middle of the second millennium before the present era, to 800 or 700 B.C.[15] These climatic and tectonic catastrophes precipitated the wandering of hordes of destitute human beings, including, after the last catastrophe, Celts and Cimbrians.[16] The migrants came to the desolate lands from other, faraway regions, probably equally fearfully devastated.

Dropped Ocean Level

In many places of the world the seacoast shows either submerged or raised beaches. The previous surf line is seen on the rock of raised beaches; where the coast became submerged, the earlier water line is found chiselled by the surf in the rock below the present level of the sea. Some beaches were raised to a height of many hundred feet, as in the case of the Pacific coast of Chile, where Charles Darwin observed that the beach must have risen 1300 feet only recently—"within the period during which upraised shells have remained undecayed on the surface." He thought also that the "most probable" explanation would be that the coast level, with "whole and perfectly preserved shells," was "at one blow uplifted above the future reach of the sea," following an earthquake.[1] In the Hawaiian Islands there is a

[15] Cf. ibid., p. 295.   [16] Ibid., p. 187.
[1] Darwin, *Geological Observations on the Volcanic Islands and Parts of South America*, Pt. II, Chaps. IX and XV.

1200-foot raised beach. On Espíritu Santo Island in the New Hebrides in the southern Pacific, corals are found 1200 feet above sea level.[2] Corals do not grow high above the sea or in the depths of the sea; their formation is limited to levels close to the surface of the sea. Thus corals of bygone ages are recorders of previous sea levels.

In numerous instances evidences of submergence and emergence are seen on the same rock. One such case we have discussed—the Rock of Gibraltar. To a lesser degree the phenomenon is repeated in Bermuda. From the evidence of submerged caves, the sea level at Bermuda "must at one time have stood at least 60 to 100 feet lower than at present," while from raised beaches "it appears to have stood at one time at least 25 feet higher than at present" (H. B. Moore).

These changes date from different ages, but common to all of them is the absence of intermediate surf lines; if the emergence or submergence had been gradual, intermediate surf lines would be seen in the rock.

R. A. Daly observed that in a great many places all around the world there is a uniform emergence of the shore line of eighteen to twenty feet. In the southwest Pacific, on the islands of Tutuila, Tau, and Ofu and on Rose atoll, all belonging to the Samoan group but spread over two hundred miles, the same emergence is evident. In Daly's opinion this uniformity indicates that the rise was due to "something else than crustal warping." A force pushing from inside would not be "so uniform throughout a stretch 200 miles long."[3] Nearly halfway round the world, at St Helena in the South Atlantic, the lava is punctuated by dry sea caves, the floors of which are covered with water-worn pebbles, "now dusty because untouched by the surf." The emergence here is also twenty feet. At the Cape of Good Hope caves and benches "also prove recent and sensibly uniform emergence to the extent of about 20 feet."

Daly proceeds: "Marine terraces, indicating similar emergence, are found along the Atlantic coast from New York to the Gulf of Mexico; for at least 1000 miles along the coast of eastern Australia; along the coasts of Brazil, southwest Africa,

[2] L. Don Leet, *Causes of Catastrophes* (1948), p. 186.
[3] Daly, *Our Mobile Earth*, p. 177.

and many islands of the Pacific, Atlantic, and Indian Oceans; in all these and other published cases, the emergence is recent as well as of the same order of magnitude. Judging from the condition of benches, terraces, and caves, the emergence seems to have been simultaneous on every shore."[4]

Of course Daly also found many places where the change in the position of the shore line was of a different magnitude, but "these local exceptions prove the rule." In his opinion, the cause of the world-wide emergence of the shore lies in the sinking of the level of all seas on the globe, "a recent world-wide sinking of ocean level," which could have been caused by water being drawn from the oceans to build the icecaps of Antarctica and Greenland. Alternatively, Daly thinks it could also have resulted from a deepening of the oceans or from an increase in their areas.

P. H. Kuenen of Leyden University, in his *Marine Geology*, finds Daly's claim confirmed: "In thirty-odd years following Daly's first paper many further instances have been recorded by a number of investigators the world over, so that this recent shift is now well established."[5]

Whatever was the cause of the phenomenon observed, it was not the result of a slow change; in such case we would have intermediate shore lines between the present surf line and the twenty-foot line on the same beaches, but there are none.

Of special interest is the time of the change. According to Daly, "This increase of the ice-cap or caps has been tentatively referred to late-Neolithic time, about 3500 years ago. At that approximate date there was some chilling of the northern hemisphere at least, following a prolonged period when the world climate was distinctly warmer than now. Late-Neolithic man lived in Europe 3500 years back."[6]

As to the date of the sudden drop of oceanic level, Kuenen writes: "The time of the movement was estimated by Daly to be probably some 3000 to 4000 years ago. Detailed field work in the Netherlands and in eastern England has shown a recent eustatic depression of the same order of magnitude as deduced by Daly. Here the time can be fixed as roughly 3000 to 3500 years ago."[7] Thus the work in the Netherlands and in England

[4] Ibid., p. 178. [5] P. H. Kuenen, *Marine Geology* (1950), p. 538.
[6] Daly, *Our Mobile Earth*, p. 179. [7] Kuenen, *Marine Geology*, p. 538.

confirmed not only Daly's finding but also his dating. The ocean level dropped, of course, all over the world. It was not a slow subsidence of the bottom, or a slow spread of the ocean over land, or a slow evaporation of oceanic water: whatever it was, it was sudden and therefore catastrophic.

Thirty-five hundred years ago was the middle of the second millennium before the present era, at the close of the Middle Bronze Age in Egypt.

The North Sea

The stormy North Sea, bordered by Scotland, England, the Low Countries, Germany, Denmark, and Norway, is a very recent basin. The geologists assume that the area was once before occupied by a sea, but that early in the Ice Age the detritus carried from Scotland and Scandinavia filled it, so that there was no sea left: it was all turned into land. The river Rhine flowed through this land and the Thames was its tributary; the mouth of the river was somewhere near Aberdeen.

In post-glacial times, so it is assumed, in the Subboreal period, which began about 2000 years before the present era and endured to about 800 B.C., large parts of the area were added to the sea. The Atlantic Ocean sent its waters along the Scottish and Norwegian shores, and also through the Channel that had been formed only a short while before. Human artifacts and bones of land animals were dredged from the bottom of the North Sea; and along the shores of Scotland and England, as well as on the Dogger Bank in the middle of the sea, stumps of trees with their roots still in the ground were found. Forty-five miles from the coast, from a depth of thirty-six metres, Norfolk fishermen drew up a spearhead carved from the antler of a deer, embedded in a block of peat.[1] This artifact dates from the Mesolithic or early Neolithic Age and serves as one of many proofs that the area covered by the North Sea was a place of human habitation not many thousands of years ago. From the analysis of the pollens found in the peat taken from the bottom of the sea,

[1] E. Janssens, *Histoire ancienne de la Mer du Nord* (2nd ed.; 1946), p. 7; K. Gripp, "Die Entstehung der Nordsee," in *Werdendes Land am Meer* (1937), pp. 1–41.

the conclusion was reached that these forests existed in not too remote times. It has also been assumed that the building of large areas of the North Sea in the Subboreal period resulted from a rather sudden sinking of the land, which some authorities date at about 1500 B.C. or a little earlier, at the same time that floods destroyed the lake dwellings of central Europe.

If we consider that Phoenician vessels were already visiting the Atlantic coast of Europe in the days of the Middle Kingdom in Egypt, or before 1500 B.C., we begin to see in its historical perspective the catastrophe that spread the North Sea over inhabited land. The submerged land must have been occupied by human settlements of the Mesolithic and Neolithic ages, whereas Egypt and Phoenicia had already reached the Middle Bronze.[2]

The sea did not slowly encroach, finally to evict the population of the settlements; it entered the land without much warning, and sent its dark billows rolling to find new barriers. The Dogger Bank may have stood out for some time longer, but at last it, too, was taken over by the sea.

After a hundred generations, man began with great effort to recapture bits of land from the sea, building dams and sluices; at this work he, too, discovered bones of animals in vast and tangled masses, of extinct and living forms, generally ascribed to the Ice Age. So, in the Dutch village of Tegelen, in a layer of sand, silt, clay, and peat, ancient elm, ash, and grape were found with extinct fresh-water snails, with bones of elephants, mammoths, rhinoceroses, hippopotami, deer, horses (*Equus stenonis*), and hyenas.[3]

A recent investigation of the English Fens by H. Godwin of Cambridge University, with emphasis on the plant life in the post-glacial period, disclosed a "general transgression" of the sea "in the period between the Neolithic and Romano-British times, for which our evidence is best."[4]

---

[2] Janssens, however, writes: "*L'ouverture de la mer du Nord sur l'océan Atlantique est donc beaucoup plus récente que la coupure de la Meditérranée aux colonnes d'Hercule; elle coincide à peu près avec l'épanouissement de la civilisation sumérienne en Mésopotamie.*"

[3] Flint, *Glacial Geology and the Pleistocene Epoch*, p. 325.

[4] H. Godwin, "Studies of the post-glacial history of British vegetation," *Transactions of the Royal Society of London*, Ser. B, Vol. 230, February 1940.

The Fens occupy an area of about two thousand square miles of Lincolnshire, Cambridge, and Norfolk counties, running east of Norfolk and around the Wash, a gulf of the North Sea. "The transgression was broken by two periods of retrogression, one in the Bronze Age, and the other after [the beginning of] the Iron Age."

Within the Neolithic period "the forest trees . . . all fell to the northwest. These fallen forests were mostly oak." Along with the oak trees were found tools of polished stone. Some time after the hurricane that broke all the oaks came another calamity: the land "was now suddenly changed by an extensive invasion by the sea." "Within a short time" almost the whole of the fenland area became a brackish lagoon, which later became a fresh-water area again. Bronze tools and weapons are found in abundance in the peat.

The climate became "much worse with the change to the Iron Age at about 500 B.C."—other authors ascribe this *Klimasturz* to the eighth century. It turned both colder and wetter. The area grew quite uninhabitable, for no traces of pre-Roman Iron Age man have been found there. Then came the last intrusion of the sea.

Thus, according to Godwin's analysis, in the period between 2000 and 500 before the present era, the plain north of Cambridge was more than once invaded by the North Sea under circumstances that we would interpret as catastrophic.

In many places all around England and Wales there are submerged forests which are dated as "probably Post-Glacial or Recent."[5] On the other hand, their submersion did not take place "within the past 2500 years." Some of the submerged forests have the stumps of their trees "rooted on the spot." The list of these forests is long.[6]

Submerged forests were observed also in many other places,

[5] H. B. Woodward, *The Geology of England and Wales* (2nd ed., 1887), p. 523.

[6] Submerged forests were observed off Cardunock, on the Solway, at the Alt mouth, Great Crosby, in Poolvash Bay, at Llandrillo Bay, Cardigan Bay, St. Brides Bay, and Swansea Bay; at Holly Hazle, near Sharpness, at Stolford, near the mouth of the Parret, in Porlock Bay, in West Somerset, on the coasts of Devon, at Braunton Burrows, at Blackpool, at North and South Sands, in the Salcombe estuary, in Bigbury Bay, and in Cornwall at Looe, Fowey, Mounts Bay, and in other places. *Ibid.*, pp. 523–26.

for instance near Greenland and off the east coast of America. There exist also less reliable reports of walls of sunken cities spied under water—in the North Sea, off the Atlantic coast, in the Mediterranean, all around Europe as also in faraway places, like off the Malabar coast of India.

Only several thousand years ago, as the raised beaches and sunken forests evidence, the land rose and fell and traded its domain with the sea.

Chapter XII

THE RUINS OF THE EAST

Crete

The Isle of Crete in the blue waters of the Mediterranean, with its precipitous reddish, rocky shores, a silent monument of a world that has passed, was millennia ago a great centre of an unusually rich culture. The Minoan scripts are now in the process of being deciphered; the clue was discovered by Michael Ventris, an English architect.

The history of ancient Crete—or of the Minoan culture on it —is divided into Early, Middle, and Late Minoan ages, corresponding in time with the Old, Middle, and New Kingdoms in Egypt. The period of the Hyksos in Egypt, between the Middle and the New Kingdoms, coincides with the last—the third— subdivision of Middle Minoan.

All the great periods in Minoan Crete terminated in natural catastrophes. The monumental work of Sir Arthur Evans, *The Palace of Minos at Knossos*, furnishes abundant evidence of the physical nature of the destructive agent that brought to a close the ages of Minoan culture, one after the other. He speaks of a "great catastrophe" that took place toward the close of Middle Minoan II.[1] "A great destruction befell Knossos on the northern shore of the island and Phaestos on its southern shore."[2] The isle lay prostrate, overwhelmed by the elements.

When, finally, the survivors or their descendants began the work of restoration, their labour was destroyed again in an "overthrow."[3] Barely half a century passed between these two

---

[1] Sir Arthur Evans, *The Palace of Minos at Knossos* (1921–35), III, 14.
[2] Ibid., II, 287; III, 347.   [3] Ibid., II, 348.

catastrophes: one synchronical with the end of the Middle Kingdom in Egypt and the Exodus,[4] the other, one or two generations later.

In the later phase of Middle Minoan III the phenomena "conclusively point to a seismic cause for the great overthrow that befell the Palace and surrounding Town."[5] "Throughout the exposed areas of the building [palace] there is evidence of a great overthrow, burying with it a long succession of deposits. . . ."[6]

At the end of the next age, Late Minoan I, the existence of the palace of Knossos "was cut short by some extraneous cause, though without any such signs of wholesale ruin as seem to have marked its earlier disaster."[7] However, Martinatos, director of the Greek Archaeological Service, finds: "The catastrophe of Late Minoan I was fatal and general throughout the whole of Crete. It seems certain that it was the most terrible of all which occurred on the island." The palace at Knossos was destroyed. "The same tragedy befell all the so-called mansions. . . . Whole cities, too, were destroyed. . . . Even sacred caves fell in like the one at Arkalokhori."[8] Volcanic ash fell on the island and great tidal waves moved toward the island from the north and swept over it. In this catastrophe Crete received "an irreparable blow." The only explanation for the upheaval "is one of natural causes; a normal earthquake, however, is wholly insufficient to explain so great a disaster."[9]

Then came the destruction of Late Minoan II. The sudden catastrophe interrupted all activity; but there are indications also that, though the upheaval was instantaneous, some preparations had been made in an effort to appease the deity for fear of the impending event. Evans writes: "It would seem that preparations were on foot for some anointing ceremony. . . . But the initial task was never destined to reach its fulfilment."[10] Beneath a covering mass of earth and rubble lies the "Room of

[4] The synchronism of the end of the Middle Kingdom in Egypt and the Exodus is dealt with in *Ages in Chaos*.

[5] Evans, *The Palace of Minos*, II, 347.

[6] Ibid., p. 288. [7] Ibid., p. 347

[8] S. Martinatos, "The Volcanic Destruction of Minoan Crete," *Antiquity*, XIII (1939), 425ff.

[9] Ibid., p. 429.

[10] Evans, *The Palace of Minos*, Vol. IV, Pt. 2, p. 942.

the Throne" with alabaster oil vessels. "The sudden breaking off of tasks begun—so conspicuous . . . surely points to an instantaneous cause."[11] It was "another of those dread shocks that had again and again caused a break in the Palace history." The earthquake was accompanied by fire. The actual over-throw was greatly aggravated by "a widespread conflagration," and the catastrophe attained "special disastrous dimensions owing to a furious wind then blowing." Evans assigns the final destruction of the building to the month of March. The disaster, however, did not approach in magnitude that "which, for example, had put an end to the building in its Middle Minoan age."

After this last catastrophe the palace at Knossos was never again rebuilt.

From the topography of Knossos and its surroundings it appears that sometime in the past the site of this city was at the head of an inner harbour connected by a channel with a larger harbour the entrance to which was between two headlands to the north. "Some tremendous catastrophe had raised that sec-tion of the island far above the level which it occupied when the city of Cnossus [Knossos] existed."[12]

Archaeological work on Crete disclosed vast catastrophes of a physical nature. Since the termination of the cultural ages on Crete coincided with the end of historical periods in Egypt, also brought to their end by natural catastrophes, the extent of these repeated upheavals appears not to have been local.

The island of Crete presents excellent ground for examination of the effect of the great catastrophes of the past on an early civilization. The island was not invaded until the arrival of the Dorians, so that the effects of a natural disaster cannot be mis-taken for destruction by the hand of man.

North of Crete is the volcanic island of Thera, or Santorin. The volcano is not yet extinguished. Its crater was blown off in a formidable explosion in the past and a large caldera was formed. A German-Greek expedition explored the island and published a detailed account of the vehement explosion of the

[11] Ibid.
[12] From a written communication by Norman E. Merrill, Commander, U. S. C. G.

166

former age. At that time villages were buried by lava, pumice, and ashes; the excavated cultural remains showed that the great explosion took place "between 1800 and 1500 B.C.," or at the end of the Middle Kingdom in Egypt.[13] The erupted masses were so vast that a German scholar in recent years a theory according to which the Egyptian plague of darkness was caused by the eruption of the Thera volcano, six hundred miles northwest of the Delta.

In Egypt the rock structure of the land experienced at least localized displacements at the end of the Middle Kingdom. K. R. Lepsius observed that the Nilometers at Semneh, dating from the Middle Kingdom, show an average rise in the waters of the Nile at that place, where the river is channelled in rock, twenty-two feet higher than the highest level of today.[14] "We obtained the remarkable result that about 4000 years ago the Nile used to rise at that point, on an average, twenty-two feet higher than it does at present."

This dropping of the high-water level must be ascribed either to a change in the quantity of water in the Nile or to a change in the rock structure of Egypt. However, if the Nile contained so much more water in the past, many residences and temples would have been regularly inundated.

I omit the references to cities swallowed by the ground in Egyptian literature; yet the enigmatic and rather regular signs of fires in graves of the Old and Middle Kingdoms, as if from the presence of some volatile substance that penetrated there and became inflamed by the heated ground, is worth mentioning.

Troy

At the westernmost end of Asia Minor, a few miles from the Dardanelles, lies the village of Hissarlik. In 1873, Heinrich Schliemann, though not an archaeologist, discovered there the remains of the fortress sung in the *Iliad*. From his early years as

[13] H. Reck, ed., *Santorin* (1936), p. 82; H. S. Washington in *Bulletin of the Geological Society of America*, XXXVII (1926).
[14] Lepsius, *Letters from Egypt, Ethiopia and the Peninsula of Sinai*, (1853), pp. 19-20.

grocer's apprentice, cabin boy on a ship that was wrecked, and bookkeeper in Holland, he had nourished the ambition to find Troy. After many wanderings that took him to Russia and California and the Far East, he settled in Greece, published his prediction of where he would find the city of the *Iliad*, and was met with jeers. But he soon succeeded in locating the legendary city in the Turkish village of Hissarlik.[1] It had been built six or seven times and as many times destroyed. Schliemann took the rich city on the second lowest level to be the Troy of King Priam, which endured the siege and then succumbed to the Greeks, or Achaeans, warriors under Agamemnon. Later scholars have identified the second city as of a much earlier date, and declared the sixth city from the bottom to be that of Priam and Homer. The second city came to an end at the time the Old Kingdom of Egypt fell; it was destroyed in a violent paroxysm of nature.

The archaeological expedition of Cincinnati University under Carl Blegen has established that an earthquake destroyed the city besieged by Agamemnon.[2] Claude Schaeffer, the excavator of Ras Shamra (Ugarit) in Syria, came to Troy to compare the finds of Blegen with his own at Ras Shamra and became convinced that the earthquakes and conflagrations he had noted at Ras Shamra were synchronical with the earthquakes and conflagrations of Troy, six hundred miles away. He then compared the findings of these two places with signs of earthquakes in numerous other localities of the ancient East. After painstaking work he came to the conclusion that more than once in historical times the entire region had been shaken by prodigious earthquakes, an area unusually large when compared with the largest areas affected by earthquakes in modern times. He wrote:

"There is not for us the slightest doubt that the conflagration of Troy II corresponds to the catastrophe that made an end to the habitations of the Old Bronze Age of Alaca Huyuk, of Alisar, of Tarsus, of Tepe Hissar [in Asia Minor], and to the catastrophe that burned ancient Ugarit (II) in Syria, the city

[1] At the end of the eighteenth century, in a time before the modern era of archaeology, Le Chevalier made a guess that Hissarlik was the site of Homeric Troy, or Ilion. This early identification was neglected.
[2] C. W. Blegen, "Excavations at Troy, 1936," *American Journal of Archaeology*, XLI (1937), 35.

of Byblos that flourished under the Old Kingdom of Egypt, the contemporaneous cities of Palestine, and that was among the causes which terminated the Old Kingdom of Egypt."[3] After a time of decay most of these cities were restored in a new era of rich civilization.

The city subsequently constructed, Troy III, was also destroyed in a great and sudden catastrophe; it was "a most terrible fire." Dörpfeld, the renowned archaeologist, who worked with Schliemann and survived him by many years, wondered how a town like Troy III could have left, as a result of this fire, ashes sixteen metres (over fifty feet) thick.[4] Schaeffer found that the same destruction also spread all over Asia Minor and far beyond.

Efforts to build a new city, Troy IV, on the ashes of the old were cut short by a new and unexpected conflagration. Once again the ground was covered "with a thick bed of ashes and carbonized substance indicating clearly that the buildings fell during a fire."[5]

Troy VI, which followed the fifth city and is usually recognized as the capital of King Priam, was destroyed by an earthquake. A natural force more powerful than the army of Agamemnon brought about its end. It was a violent shaking of the ground, as is also narrated in the *Iliad*. Walls were moved from their places and fell flat. Schaeffer was once more impressed by the signs of a simultaneous upheaval in all the excavated sites of Asia Minor and the ancient East generally, and dedicated himself to collating the archaeological material of the third and second millennia before the present era with the special purpose of establishing the stratigraphic synchronism based on the sudden and simultaneous interruption of cultural ages in this entire area.

The Ruins of the East

In the ruins of excavated sites throughout all lands of the ancient East signs are seen of great destruction that only nature

[3] Claude F. A. Schaeffer, *Stratigraphie comparée et chronologie de l'Asie Occidentale (IIIe et IIe millénaires* (Oxford University Press, 1948), p. 225.
[4] Ibid., p. 237. W. Dörpfeld, *Troja und Ilion* (1902)
[5] Blegen, *American Journal of Archaeology*, 1937, pp. 570ff.

could have inflicted. Claude Schaeffer, in his great recent work, discerned six separate upheavals. All of these catastrophes of earthquakes and fire were of such encompassing extent that Asia Minor, Mesopotamia, the Caucasus, the Iranian plateau, Syria, Palestine, Cyprus, and Egypt were simultaneously overwhelmed. And some of these catastrophes were, in addition, of such violence that they closed great ages in the history of ancient civilizations.

The enumerated countries were the subject of Schaeffer's detailed inquiry; and recognizing the magnitude of the catastrophes that have no parallels in modern annals or in the concepts of seismology, he became convinced that these countries, the ancient sites of which he studied, represent only a fraction of the area that was gripped by the shocks.

The most ancient catastrophe of which Schaeffer discerned vestiges, took place between 2400 and 2300 before the present era. It spread ruin from Troy to the valley of the Nile. In it the Old Bronze Age found its end. Laid waste were cities of Anatolia, like Alaca Huyuk, Tarsus, Alisar; and those of Syria, like Ugarit, Byblos, Chagar Bazar, Tell Brak, Tepe Gawra; and of Palestine, like Beth-Shan and Ai; and of Persia, and of the Caucasus. Destroyed were the civilizations of Mesopotamia and Cyprus, and the Old Kingdom in Egypt came to an end, a great and splendid age. In all cities walls were thrown from their foundations, and the population markedly decreased. "It was an all encompassing catastrophe. Ethnic migrations were, no doubt, the consequence of the manifestation of nature. The initial and real causes must be looked for in some cataclysm over which man had no control."[1] It was sudden and simultaneous in all places investigated.

In a few centuries, migrating and multiplying themselves, the descendants of the survivors of the ruined world built new civilizations: the Midde Bronze Age. In Egypt it was the time of the Middle Kingdom, a short but glorious resurrection of Egyptian civilization and might. Literature reached its perfection, political might its apogee. Then came a shock that in a single day made of this empire a ruin, of its art debris, of its population corpses. Again it was the entire ancient East, to its

[1] This and the following quotations are from Schaeffer, *Stratigraphie comparée*, pp. 534–67.

uttermost frontiers, that fell prostrate; nature, which knows no boundaries, threw all countries into a tremor and covered the land with ashes.

"This brilliant period of the Middle Bronze Age, during which flourished the art of the Middle Kingdom in Egypt and the exquisite art and industry of the Middle Minoan Age [on Crete], and in the course of which great centres of trade like Ugarit in Syria enjoyed remarkable prosperity, was suddenly terminated. . . .

"The great activity of international trade, which, during the Middle Bronze Age, had been characteristic of the eastern Mediterranean and most of the lands of the Fertile Crescent, suddenly stopped in all this vast area. . . . In all the sites in Western Asia examined up to now a hiatus or a period of extreme poverty broke the stratigraphic and chronological sequence of the strata. . . . In most countries the population suffered great reduction in numbers; in others settled living was replaced by a nomadic existence."

In Asia Minor the end of the Middle Bronze came suddenly, and a rupture between that age and the Late Bronze is evident in "all sites that were stratigraphically examined." Troy, Boghazkoi, Tarsus, Alisar, present the same picture of life vanishing with the end of the Middle Bronze.

In Tarsus, between the strata of the "brilliantly developed civilization" of the Middle Bronze and those of the Late Bronze, a layer of earth five feet thick was found without a sign of habitation—a "hiatus." At Alaca Huyuk the transition from Middle Bronze to Late Bronze was marked by upheaval and destruction, and the same may be said of every excavated site in Asia Minor.

On the Syrian coast and in the interior "we find a stratigraphic and chronological rupture between the strata of the Middle Bronze and Late Bronze at Qalaat-er-Rouss, Tell Simiriyan, Byblos, and in the necropoles of Kafer-Djarra, Qrayé, Majdalouna." All the necropoles examined in the upper valley of the Orontes ceased to be used, and habitation of the great site of Hama was interrupted at the moment the Middle Kingdom in Egypt went down. Also in Ras Shamra there is a marked gap between the horizons of the Middle and the Late Bronze.

In Palestine, at Beth Mirsim, there was an interruption in the habitation of the site after the fall of the Middle Kingdom in Egypt. In Beth-Shan, between the layers of the Middle Bronze and Late Bronze, the excavators came upon an accumulation of debris a metre thick. "It indicates that the transition from the Middle Bronze to the Late Bronze was accompanied by an upheaval that broke the chronological and stratigraphical sequence of the site." A similar situation was found at Tell el Hésy by Bliss. Earth tremors played havoc also with Jericho, Megiddo, Beth-Shemesh, Lachish, Ascalon, Tell Taanak. The excavators of Jericho found that the city had been repeatedly destroyed. The great wall surrounding it fell in an earthquake shortly after the end of the Middle Kingdom.[2]

Concussions devastated the entire land of the Double Stream. The Russian-Persian borderland also shows that there was no continuity between the Middle Bronze and Late Bronze. In the Caucasus not an archaeological vestige was found of the centuries between these two ages.

A sea tide broke onto the land, as on the coast of Ras Shamra, bringing further destruction in its wake.

It appears also that the end of the Middle Kingdom was marked by volcanic eruptions and lava flows. On the Sinai Peninsula, at an early and undisclosed date, a flow of basaltic lava from the fissured ground—the Sinai massif is not a volcano —burned down forests, leaving a desert behind.[3] In Palestine lava erupted, filling the Jezreel Valley. Early in this century a Phoenician vase was found imbedded in lava. Geologists have asserted that volcanic activity in Palestine ceased in prehistoric times. "The assertion of the geologists thus becomes very questionable," wrote an author at that time.[4] The vase found in lava proves volcanic activity there "in historical times." The verdict of the archaeologists is that the vase "dates from the fifteenth century before the present era," and thus the eruption must have taken place in the middle of the second millennium.[5]

Egypt, according to Schaeffer, was conquered by the

[2] J. Garstang and G. B. E. Garstang, *The Story of Jericho* (1940).
[3] Flinders Petrie, "The Metals in Egypt," *Ancient Egypt*, 1915.
[4] H. Gressmann, *Palästinas Erdgeruch in der Israelitischen Religion* (1909), pp. 74–75.
[5] Ibid., p. 75; A. Lods, *Israel* (1932), p. 31; I. Benzinger, *Hebräische Archaeologie* (3rd ed.; 1927).

Hyksos, coming from the East, when it fell in a catastrophe caused by natural elements. In other countries, too, not conquerors or migrating hordes but earthquakes and fire were the agents of destruction. "Our inquiry has demonstrated that these repeated crises which opened and closed the principal periods of the third and second millennia were caused not by the action of man. Far from it, because compared with the vastness of these all-embracing crises and their profound effects, the exploits of conquerors . . . would appear only insignificant."[6]

Schaeffer finds indications that the climate changed abruptly in the wake of the catastrophes; the phenomenon was ubiquitous: "At the same time in the Caucasus and in certain areas of prehistoric Europe, changes of climate have caused, as it appears, transformations in the occupation and economy of the countries."[7]

The catastrophe that served as the starting point for two of my works, *Worlds in Collision* and *Ages in Chaos*, left archaeological imprints on biblical and Homeric lands, from the Dardanelles to the Caucasian barrier, the Persian highland, and the cataracts of the Nile. The most severe and devastating upheaval took place exactly at the end of the Middle Kingdom in Egypt, as claimed in these two books.

What was the nature of the perturbations that caused the end of the old Bronze Age and then of the Middle Bronze Age, and changed the entire aspect of the known world from Europe to Asia and Africa? Fire raged, lava flowed, tremors travelled across whole continents, and climate went through revolutions. Schaeffer wondered at the vast extent of the earthquakes, unknown in modern annals. He asked: Could it be that in earlier times earthquakes were of very much greater force and wider spread than they are now because geological strata, originally out of equilibrium, were settling with the passing of time?[8] This explanation of the readjustment of geological strata as time goes on is not valid if we keep in mind that geology ascribes to this planet three billion years of existence, and three thousand years is just one millionth of this period. The earth would have

[6] Schaeffer, *Stratigraphie comparée*, p. 565.
[7] Ibid., p. 556.
[8] Ibid., Avant-propos, p. xii.

173

adjusted its strata long before, in the geological ages. Apparently the earth was thrown out of equilibrium only a few thousand years ago, which also explains the change in climate simultaneous with the upheaval.

Schaeffer's investigation reaches Persia in the East; inquiring in lands beyond Persia, we find that a rich Indus Valley civilization, with many fortified cities, came to a sudden end in the fifteenth century before the present era, shortly before the arrival of the Aryans. The cause of this sudden termination, "conveniently equated with the fifteenth century B.C.," is not known; but the facts brought forth by R. E. Mortimer Wheeler[9] strongly suggest to various scholars[10] that a natural catastrophe engulfed the area in those early Vedic times. In its wake the Aryans came into the country; a Vedic Dark Age ensued, and on the ashes of the effaced world Aryans, step by step, built a new civilization.

Times and Dates

The evidence of this and preceding chapters should not be interpreted as proving that there were global catastrophes only in the first and second millennia before the present era; but as substantiating the claim that in those times, too, there were global disturbances: these were actually the last in a line that goes back to much earlier times.

According to the narrative of *Worlds in Collision*, two series of world catastrophes took place in recent times: "one that occurred thirty-four to thirty-five centuries ago, in the middle of the second millennium before the present era; the other in the eighth and the beginning of the seventh century before the present era, twenty-six centuries ago."[1] The first of these catastrophes occurred at the end of the Middle Kingdom in Egypt and actually caused its termination; in *Ages in Chaos* further details were given of the closing hours of the Middle

[9] R. E. Mortimer Wheeler, "Archaeology in India and Pakistan since 1944," *Journal of the Royal Society of Arts*, XCIX (December 1950); idem, *Pakistan, Geological Review*, Vol. I, Pt. I.

[10] A written communication of H. K. Trevaskis, author of *The Land of the Five Rivers* (Oxford University Press, 1928).

[1] *Worlds in Collision*, Preface.

Kingdom, which went down under the blows of nature. The second series of catastrophes occurred in the period that started in 776 B.C. and lasted until 687 B.C., when, in the final act of a protracted drama, Sennacherib met his downfall.

In an independent investigation, Claude Schaeffer came to the conclusion that at the end of the Middle Kingdom an enormous cataclysm took place that ruined Egypt and devastated by earthquake and holocaust every populated place in Palestine, Syria, Cyprus, Mesopotamia, Asia Minor, the Caucasus, and Persia;[2] previously Sir Arthur Evans had shown that, at the downfall of the Middle Kingdom in Egypt, Crete was overwhelmed by a natural upheaval; also the volcano of Thera erupted prodigious quantities of lava; and the Indus Valley civilization came abruptly to an end.

More recent catastrophes embracing the entire Near and Middle East are also described by Schaeffer as having taken place a few centuries later. Evans had found that the cities of Crete were again destroyed in very severe earthquakes that terminated the consecutive Minoan ages on Crete.

Schaeffer's findings, based on excavations in scores, if not hundreds, of places all around the ancient East, where populations were decimated or annihilated, the earth shook, the sea irrupted, and the climate changed, are by themselves sufficient support for the claims made in *Worlds in Collision* as to the times and the vastness of the catastrophes. But we have much more evidence, and no wonder: the catastrophes being ubiquitous, their effect must be found everywhere.

The Rhone Glacier in the Alps started to melt 2400 years ago, in the middle of the first millennium. This calculation of De Lapparent coincides with that at which we arrived by dating the last catastrophe in 687 B.C. In this catastrophe many older glaciers melted, and the subsequent increased evaporation and precipitation built other glaciers that before long also started to melt, a process that has been going on ever since. Many glaciers

---

[2] Schaeffer, in accordance with the accepted chronology, placed the end of the Middle Kingdom between 1750 B.C. and 1650 B.C. He wrote, however: "La valeur des dates absolues adoptées par nous dépend, bien entendu, pour une part, du degré de précision obtenu dans le domaine des recherches sur les documents historiques utilisables pour la chronologie. . . ." (*Stratigraphie Comparée*, p. 566). In *Ages in Chaos* I have shown why the end of the Middle Kingdom must be dated about 1500 B.C.

of the Alps, it was recently learned with surprise, are less than 4000 years old (Flint).

Catastrophic changes in climate, found by Sernander and others in Scandinavia, correspond almost exactly with our dates: in the second millennium, about 1500 B.C. and once more, 800 to 700 years before the present era, or thirty-four and almost twenty-seven centuries ago. The same dates are established through pollen analysis by Gams and Nordhagen for the catastrophic changes of climate in German fens and tectonic disturbances in central Europe; and again the same dates, close to the middle of the second millennium before the present era and once more following the year 800 B.C. are fixed by Paret and other authors for the climatic catastrophes that are reflected in the history of the lake dwellings in Germany, Switzerland, and northern Italy.

Careful investigation by W. A. Johnston of the Niagara River bed disclosed that the present channel was cut by the falls less than 4000 years ago. An equally careful investigation of the Bear River delta by Hanson, who compared measurements repeated in periodic surveys, showed that the age of this delta is 3600 years, its origin going back to the middle of the second millennium before the present era.

Warren Upham's research on the great glacial Lake Agassiz and the striations of the exposed rocks there indicates that the lake was formed but a few thousand years ago and existed for a short time only.

The study by Claude Jones of the lakes of the Great Basin showed that these lakes, remnants of larger glacial lakes, have existed only about 3500 years, and also that the Ice Age fauna survived to a date equally recent. Gale obtained the same result on Owens Lake in California and also Van Winkle on Abert and Summer lakes in Oregon.

Radiocarbon analysis by Libby also indicates that plants associated with extinct animals (mastodons) in Mexico are probably only 3500 years old. Similar conclusions concerning the late survival of the Pleistocene fauna were drawn by various field workers in many parts of the American continent.

Suess and Rubin found with the help of radiocarbon analysis that in the mountains of the western United States ice advanced only 3000 years ago.

The study of the magnetic properties of the clay of Etruscan vases points to a reversal of the general magnetic field of the earth, and also to a passage of the earth through strong magnetic fields in historical times.

The Florida fossil beds at Vero and Melbourne proved—by the artifacts found there together with human bones and the remains of animals, many of which are extinct—that these fossil beds were deposited between 2000 and 4000 years ago. As brought out by Godwin, the two irruptions of the sea on English shores also took place sometime in the second and first millennia before the present era. According to an earlier work, by Prestwich, the irruption of the sea was of a very violent nature; it spread to central France and the French Riviera, to Gibraltar, Corsica, and Sicily, and to the entire area that stretches to the lands of the ancient East. In all these places animal bones have been found broken but fresh; these bones of extant and extinct species have been found in fissures and caverns, sometimes on the tops of high hills, in great numbers. The bones found in English caves, covered with diluvium, were also described as fresh and unfossilized.

From observations on beaches in numerous places all over the world, Daly concluded that there was a change in the ocean level, which dropped sixteen to twenty feet 3500 years ago; Kuenen and others confirmed Daly's findings with evidence derived from Europe.

To these closely dated geological, climatological, and archaeological evidences of catastrophes, we may add numerous others which also point to the recency of the great upheavals.

Animals, torn and broken, many of which are of extinct forms, are found in enormous heaps in Alaska, their bones and skin still fresh; the mammoth meat discovered in Siberia is still edible; the bones of hippopotami in the rock fissures of England still retain their organic matter. The mountain chains of China and Tibet, of the Andes, the Alps, the Rockies, and the Caucasus rose to their present heights in the Late Stone and even in the Bronze Age, and at those times (post-glacial) Africa was torn by the Great Rift.[3]

We have the same late dating from all parts of the world, and what is even more important, by all kinds of calendars,

[3] See page 82.

calculations, and approaches. And actually the figures brought together on these pages are from the fields of archaeology and climatology, and from fossil beds and waterfalls and deltas and fens (pollen analysis), from lake dwellings and glaciers and ocean levels and the magnetic polarity of the earth, disclosing the same events and the same dates.

Chapter XIII

COLLAPSING SCHEMES

Geology and Archaeology

Measured by anthropological and archaeological evidence, the age of many finds is recent; measured by the prevailing geological and paleontological schemes, the dates of the same finds are many times more remote. This conflict was very sharp in the case of the Vero and Melbourne, Florida, beds containing fossils and artifacts, and it repeated itself in a great many places. A. S. Romer brought together a wealth of material to show the late survival of Pleistocene fauna and was widely quoted by archaeologists. A. L. Kroeber sees no easy way to avoid the conclusion that "some of the associations of human artifacts with extinct animals may be no more than three thousand years old" and not "twenty-five thousand years old."[1] Like Jones, he assumes that the Ice Age fauna survived until such a recent time by going through a process of slow extinction. But the idea of the slow and gradual extinction of Ice Age fauna is opposed by students of the problem, who feel that "sudden and decisive geological or climatic changes occurred which simultaneously wiped out a considerable number of animal species."[2]

From the evidence turned up on the European continent, "where documentation from early post-glacial sites is much more complete, we find a rather sudden disappearance" of the fauna.[3]

When measured by archaeological standards, however, the artifacts and other remains of human origin found with the

[1] A. L. Kroeber in the volume dedicated to A. M. Tozzer, *The Maya and Their Neighbours* (1940), p. 476.

[2] L. C. Eiseley, "Archaeological Observations of the Problem of Post-Glacial Extinction," *American Antiquity*, Vol. VIII, No. 3 (1943), p. 210.

[3] Ibid., p. 211.

fossils point to a much closer date in Europe too. K. S. Sandford, writing of the conflict of views between geologists and archaeologists in England, says: "The difference of opinion in some instances is so complete that one or the other must assuredly be wrong."[4] Those who measure the time in terms of cultural or physical anthropology and archaeology stand in very definite opposition to all estimates based on a geological or a paleontological time scale.

As an additional argument the archaeologist points to pictures of extinct animals in Babylonian and Egyptian bas-reliefs, the bones of which have actually been found. And the anthropologist believes that even oral traditions concerning extinct animals are grounds for far-reaching conclusions.

"Archaeology has proved that the American Indian hunted and killed elephants; it has also strongly indicated that these elephants have been extinct for several thousand years. This means that the traditions of the Indians recalling these animals have retained their historical validity for great stretches of time. Exactly how long, it is impossible to say: probably the minimum is three thousand years. . . . If some Indian traditions have remained historical for so many years, undoubtedly traditions of other races and peoples have also."[5]

The animals of the La Brea asphalt pits in Los Angeles were first regarded as belonging to the opening of the Pleistocene or Ice Age, almost a million years ago; then, the close relation between the Lahontan fossils and those of La Brea compelled a change in this estimate and the assignment of the fauna of La Brea, as well as the similar fauna of other asphalt pits in California (Carpinteria and McKittrick) to the end of the Ice Age, presumably twenty or thirty thousand years ago.

"Perhaps most striking is the conclusion that if these so-called early Pleistocene assemblages are actually late Pleistocene in age, early Quaternary vertebrate faunas are as yet practically unknown in the western United States."[6]

[4] K. S. Sandford, "The Quaternary Glaciation of England and Wales," *Nature*, December 2, 1933.

[5] L. H. Johnson, "Men and Elephants in America," *Scientific Monthly*, October 1952.

[6] J. R. Schultz, "A Late Quaternary Mammal Fauna from the Tar Seeps of McKittrick, California," in *Studies on Cenozoic Vertebrates of Western North America* (Carnegie Institution, 1938).

This radically revised view was not limited to the western coast of North America: the fauna that, two or three decades ago, was thought to have perished at the advent of the glacial periods is now thought to have survived the entire Ice Age and to have perished at its very end.

"It seems odd that a fauna which had survived the great ice movement should die at its close. But die it did."[7]

Yet even the reduction of the time when the major part of the Pleistocene fauna succumbed on the western coast from one million years to only thirty or twenty or even ten thousand years is insufficient if Jones' estimate of the age of Lahontan deposits is correct. According to his analysis of the salt accumulation in the residual lakes of the larger Lake Lahontan, this glacial lake came into existence only 3500 years ago, and the fauna found in its deposit could not be older. This compelled further vacillations. J. R. Schultz, writing on the fauna of the tar seeps in California, says that in view of the established correlation of the fauna of La Brea and the fauna of Lake Lahontan it is now possible "to reconcile the vertebrate evidence" even with the opinion of Jones "as to the relatively late age of the lake."[8] Would this really signify that the extinct animals of the asphalt pits are only 3000 or 4000 years old? This would mean that these bones were deposited in the time of the recorded history of Egypt and Babylonia.

Thus we witness a return to the view held by American geologists in the latter part of the nineteenth and the beginning of the present centuries: George Frederick Wright (1838–1921), Newton Horace Winchell (1839–1914), Warren Upham (1850–1934). Wright concluded that the Ice Age "did not close until about the time that the civilization of Egypt, Babylonia and Western Turkestan had attained a high degree of development," and this in opposition to the "greatly exaggerated ideas of the antiquity of the glacial epoch."[9]

Toward this view, with slow steps, scientific opinion is approaching, though it still maintains that there was a great gap between the Ice Age and the beginning of recorded history, the

[7] Eiseley, *American Antiquity*, Vol. VIII, No. 3 (1943), p. 211.
[8] Schultz, in *Studies on Cenozoic Vertebrates*.
[9] Wright, *The Ice Age in North America*, p. 683.

survival of many Ice Age animals until the second millennium before the present era notwithstanding.

Collapsing Schemes

In 1829 Gérard Deshayes published his studies on the fossiliferous strata in the Paris area, where marine animals alternate with land animals; these strata disclosed that in the upper marine bed were many kinds of shell-bearing molluscs that still inhabit the waters of the sea, and that the deeper the stratum, the fewer the living forms of molluscs.

Following the publication of Deshayes's work, Lyell devised a timetable of geological ages. The fossilized remains of ancient animals indicate changes in fauna in the course of time; Lyell's measurement of geological periods is based on such changes in the animal kingdom, especially among the shell fauna. He found that there has been in the Quaternary, or the age of man, not more than one twentieth of the evolution that has occurred since the lower Miocene (middle series of Tertiary, the age of mammals). From that point on he traced one complete "cycle of evolution," during which, at his estimate, practically all species that existed at the opening of the cycle were replaced by new species. Thus, if a figure of 1,000,000 years is accepted for the age of man, which started with the close of the Tertiary epoch, then 20,000,000 years were needed to accomplish the changes observed since the lower Miocene; and four such cycles of transformation of life must have passed since the end of the Mesozoic, or the age of reptiles. By this method Lyell reckoned twelve cycles, or 240,000,000 years, from the beginning of the Paleozoic, or the time of early life forms on the earth. This figure is now considerably increased; the other figures are accepted at Lyell's valuation.

Lyell's scheme, perfected by the introduction of new subdivisions of geological epochs, sets forth the following rule. If a stratum contains ninety to a hundred per cent modern species of shells, the stratum is Pleistocene, or of the Ice Age; if it contains forty to ninety per cent modern species of shells, the stratum belongs to the last subdivision of Tertiary—the Pliocene; if only twenty to forty per cent of the shells in a stratum are present-day varieties, then the stratum is of Miocene time, an earlier subdivision of the Tertiary; and so on, down to the stratum where shells of extant species of molluscs find no direct ancestors.

Lyell's time system is based on the assumption that no catastrophic events intervened and that the extirpation of species was the result of slow extinction, which Darwin's theory ascribes to the survival of the fittest in the struggle for the limited means of existence. But if great catastrophes occurred on the surface of the earth and in the depths of the seas, of more than local character, and if in such upheavals some forms of life perished and others survived, and the progeny of still others underwent strong mutations, then the entire scheme of percentages and time allotment by the multiplication of changes observed in the last epoch, with its preconceived plan and rigidity, is no more valid than the pronouncements of some theologians, like Archbishop Ussher of Ireland, who in 1654 declared that the Creation took place at nine o'clock in the morning on the twenty-sixth day of October in the year 4004 B.C.

The present work does not suggest either a lengthening or a shortening of the estimated age of the earth or the universe (which during the few years when this book was being written rose from two to six billion years). I do not see why to a truly religious mind a small and short-lived universe is a better proof of its having been devised by an absolute intelligence. Neither do I see how by removing many unsolved problems in geology to very remote ages we contribute to their solution or elucidate their enigmatic nature.

Whatever the age of the universe and the earth, single geological epochs were of very different length than has been assumed on the ground of the theory of uniformity. The very concept of a 60,000,000-year-long Tertiary when mountains were uplifted, followed by 1,000,000 years of Ice Age, a time of great climatic changes, followed by 30,000 years of the tranquillity of Recent time, with quietude in mountain building and stability in climate, is basically wrong. The mountain building went on during the Ice Age, coinciding with climatic catastrophes, and both endured into Recent time, only a few thousand years ago.

183

The Accepted Sequence of Geological Ages

CENOZOIC	Age of Man	QUATERNARY	Recent (Neolithic, Bronze, Iron)
			Pleistocene or Glacial (Paleolithic)
	Age of Mammals	TERTIARY	Pliocene
			Miocene
			Oligocene
			Eocene
MESOZOIC	Age of Reptiles		Cretaceous
			Jurassic
			Triassic
PALEOZOIC	Age of Invertebrates, Fishes, Amphibians		Permian
			Carboniferous
			Devonian
			Silurian
			Ordovician
			Cambrian
AZOIC	No life and primitive life		Pre-Cambrian

When the earlier rocks are investigated they are found to be records of great upheavals in comparison with which the up-heavals of later times appear only minor. Along the Canadian border west of Lake Superior in the Keewatin area, a complex of ancient lava flows and interbedded sedimentary rock reached, according to C. O. Dunbar of Yale, "the impressive thickness of 20,000 feet."[1] At Michipicoten Bay the volcanic tuff is 11,000 feet thick. In the same area of Lake Superior a later flow of (Keweenawan) lava, still very early in the history of the world, "has been estimated at 24,000 cubic miles," and in northern Michigan and Wisconsin, the Keweenawan system "may reach 50,000 feet, much more than half of which is made of lava flows." "It stirs the imagination to contemplate the 2,000,000 square miles of granite gneiss that floors the Canadian Shield, and to realize that it all came into place as fluid magma, which congealed beneath a cover of older rocks now long since removed by erosion." The impression is gained "that during these primeval eras the crust of the Earth was repeatedly broken and largely engulfed in upwellings of molten material." In these pre-Cambrian lavas, glacial deposits were found in Canada as well as in Australia and South Africa, "with boulders in part rounded and in part angular, and some of them faceted and striated." The detection of this evidence of early glaciation came at first as "a shocking discovery," because it appeared "a serious obstacle to the belief that the Earth was originally molten." Later, however, geologists, by placing some half a billion years between the origin of the earth and the early ice phenomena, allowed the rocks to cool off first.

Then in Cambrian time seas flooded the continents, and dolomite and metamorphosed rocks 3000 to 4000 feet thick were formed. Only lower animal life was present in the world. Yet "the simplest, unspecialized ancestors of modern animals, are most intensely modern themselves in the zoological sense and . . . belong to the same order of nature as that which prevails at the present day." In Ordovician time the sea submerged "fully half of the present [American] continent and reduced it

[1] This and the following quotations are from Dunbar, *Historical Geology* (1949); in the earlier editions Charles Schuchert figures as co-author.

to a group of great islands." In the beginning of that period, "the marine waters also spilled over and at times spread widely across the central and eastern part of the United States." Later in that period "a vast sea spread southward from the Arctic across central Canada to join the southern embayments that occupied much of the United States." Mountains were rising, folding, and overthrusting, in the so-called Taconian disturbance. This was accompanied by volcanic activity. Ash fell from Alabama to New York, "and even as far west as Wisconsin, Minnesota, and Iowa." The ash beds vary in thickness from a few inches up to more than seven feet. "The greatest display of volcanic activity, however, is found farther to the northeast, in Quebec and Newfoundland," where volcanic tuff of great thickness represents the epoch. At the same time coral reefs were built in arctic Canada, from Alaska to Manitoba, as well as in Newfoundland and northern Greenland. Indications of an ice age (tillites) are found in northern Norway, and if they are of the same age they certainly present a problem, because of the coral reefs that then grew in the north. Life was concentrated in water; the sea was inhabited by thousands of species.

In the following Silurian period volcanic activity broke out with new vigour. "In New Brunswick and especially in southeastern Maine, ash beds and lava flows attain the impressive thickness of 10,000 feet and more." Also in southern Alaska and northern California there are imposing lava flows, volcanic breccia, and tuff dating from this time. The close of this period was marked by the so-called Caledonian disturbance in Europe, with a mountain crest rising across the British Isles and Scandinavia. "Throughout the length of Norway and Sweden, a distance exceeding 1100 miles, the pre-Devonian formations were folded, overturned, and overthrust with eastward movement on individual fault planes as great as 20 to 40 miles." Again coral grew in arctic regions.

The next (Devonian) period was marked by a so-called Acadian disturbance, with uplifts and depressions. "Much igneous activity accompanied the Acadian disturbance. Great thickness of bedded lavas and tuffs in southern Quebec, Gaspé, New Brunswick, and Maine record volcanoes that were active during Devonian time." Magma intruded and lifted the White Mountains and built their granite core. Similar processes went on in other parts of the world. The Old Red Sandstone of Europe is a Devonian formation. In eastern Australia mountains were formed that stretched the full length of the eastern border of the continent. "Much igneous activity had occurred during the period in this region, and the Devonian strata and associated volcanics are said to be over 30,000 feet thick." Throughout Devonian time North America must have been connected with Europe by a land bridge "which later subsided beneath the north Atlantic." Evidence that these two lands met is found in the land plants and fresh-water animals preserved in the Devonian rocks of the two regions, "which are so much alike on both sides of the Atlantic that it seems clear they were free to migrate across an easy land bridge."

In the Carboniferous period mountains were made, seas invaded land, corals built reefs on the arctic coast of Alaska and on the polar islands of Spitsbergen, volcanoes erupted, and glaciation took place, especially in Australia. Land animals left their records beside those of rich marine life. Coal beds were formed. In the coal basins of Nova Scotia and New Brunswick, "the coal measures reach a thickness of a few thousand to 13,000 feet." Extensive continental glaciation of India, South Africa, South America, and Australia took place. . . .

Here I stop quoting from *Historical Geology*. Again, and again the world was a playground for Vulcan and Poseidon, the elementary forces of melting rock and trespassing sea. But when all is told, we are nevertheless assured that the geological record is one of calm and uniformity, and what appears as revolution is a telescoped view of slow and ordinary processes; even the seas of lava, though obviously formed in single paroxysms, are, in the over-all picture, denied a catastrophic origin.

One reads, "It is not obvious that the city of Boston rests on the surface of one of the world's greatest mountain chains—yet it does" (it had been depressed and also eroded) (Daly[2]); one reads also that "Boston lay in the equatorial rain zone during the Carboniferous and in the region of hot deserts during the Permian" (Brooks[3]); one is, furthermore, told that the site of Boston was once under the sea, and that it was once also under a mile-thick cap of ice. It is insisted that all these changes took

[2] Daly, *Our Mobile Earth*, p. 239.
[3] Brooks, *Climate through the Ages* p. 232.

place without any upheavals in nature, merely as effects of processes and agents active also in our own time—the highest mountains becoming flat, equatorial jungles giving place to hot sand deserts and hot deserts to a polar cover of ice, and the polar cover of ice to the bottom of the sea and the bottom of the sea to the site of Harvard University. It all happened so slowly that no living creature ever perceived the change.

Coal

Coal is found in layers that are ascribed to various ages mainly on the basis of fossils found in them. Brown coal is a compacted mass of plant remains. Lignite is made chiefly out of trees only partially converted into coal. Soft or bituminous coal is brittle and of bright lustre and contains sulphur; its organic nature can sometimes be seen under a lens, and the plants that participated in its formation can be recognized by leaves in the shales on top of the coal bed. Anthracite or hard coal is metamorphosed bituminous coal.

The plants that went into the formation of ancient beds include chiefly ferns and cycads; layers of later ages are composed of sassafras, laurel, tulip tree, magnolia, cinnamon, sequoia, poplar, willow, maple, birch, chestnut, alder, beech, elm, palm, fig tree, cypress, oak, rose, plum, almond, myrtle, acacia, and many other species.[1]

The origin of the coal beds is still far from being satisfactorily explained.[2] One theory would make peat bogs the place where in a slow process measured by tens and hundreds of thousands of years, coal was born. It is said that the plants fall, but before they decompose in the air they are covered by the water of the swamps. A layer of sand is deposited over them, forming the soil for new plants, and thus the process repeats itself. In order that the layer of sand may be deposited, it is necessary that these marshy regions be covered by water in motion. Since almost regularly marine shells and fossils are found on top of coal beds, the sea must have covered the swamps at one time; then, for new land plants to grow there, the sea must have retreated.

[1] Price, *The New Geology*, pp. 468–69.
[2] See Suess, *The Face of the Earth*, II, 244.

There are places where sixty, eighty, and a hundred and more successive beds of coal have formed; this theory would then require that as many times the sea trespassed—when the land slowly subsided—and as many times retreated. In other words, this theory assumes that the ground is pulsating and that the sea will return again sometime and cover the coal beds as it did a hundred times in the past.

"Fossils of marine clams, snails . . . are abundant in the shales just above each seam of coal. Later, with fluctuating sea level, the salt waters withdrew and another freshwater marsh came into being, giving rise to another bed of coal above the earlier one. Again we are surprised, this time by the large number of such alterations of coal with marine sediments; these are now recognized as distinct cycles, each cycle representing a common sequence of events. . . . Ohio displays more than forty such cycles, and in Wales more than a hundred separate seams of coal have been discovered. Marvin Miller has given 400,000 years as the probable time represented by the average Ohio cycle."[3]

This scheme demands not only that the sea should have covered the land one hundred times but also that after each retreat of the sea a fresh-water marsh should have appeared on the vacant ground in order to give the trees a place to grow and fall down and decay; and that the process of decay should have been checked before going too far, "for otherwise the vegetable matter would have disappeared completely and none would have been left in the form of coal."[4] And then each time "not only was the areal extent of the marshes remarkable but the thickness of the coal required a surprising accumulation of vegetable matter."

Many kinds of plants and trees that went into the formation of coal do not grow in swamps, and when they die they remain on dry ground and decompose. This fact suffices to render the peat-bog theory untenable.

Seams of coal are sometimes fifty or more feet thick. No forest could make such a layer of coal; it is estimated that it would take a twelve-foot layer of peat deposit to make a layer of coal one foot thick; and twelve feet of peat deposit would require

[3] Chamberlin, in *The World and Man*, ed. Moulton, p. 79.
[4] Ibid., p. 78.

plant remains a hundred and twenty feet high. How tall and thick must a forest be, then, in order to create a seam of coal not one foot thick but fifty? The plant remains must be six thousand feet thick. In some places there must have been fifty to a hundred successive huge forests, one replacing the other, since so many seams of coal are formed. But it is further questionable whether the forests grew one on top of the other, because a coal bed, undivided on one side, sometimes splits on the other side into numerous beds, with layers of limestone or other formations between.

The consideration of the enormous mass of organic matter needed to form a coal seam brought about the birth of another theory of the origin of coal. Fallen trees were carried along by overflowing rivers and coal was formed from them, not from the plants *in situ*. This theory explains the enormous accumulation of dying plants in some localities; it may be able to show why, in many cases, a fossilized tree trunk is embedded in coal with its lower part uppermost, or standing on its head—which the peat-bog theory does not explain. But the drift theory cannot account for the fact that various kinds of marine life are mixed with the coal. Carbonaceous and bituminous shales are frequently packed with fossilized marine fish. Deep-sea crinoids and clear-water ocean corals often alternate with the coal beds.

Erratic boulders, too, are often encased in coal. It was supposed that these boulders were carried by chance on natural rafts of closely drifting logs and thus became embedded in the coal. Close rafts of drifting trunks are conceivable only after a great hurricane. However, marine fish would not enter deeply into inundating rivers to be entombed together with the boulders, and coral does not grow in muddy water.

Apparently the coal was not formed in the ways described. Forests burned, a hurricane uprooted them, and a tidal wave or succession of tidal waves coming from the sea fell upon the charred and splintered trees and swept them into great heaps, tossed by billows, and covered them with marine sand, pebbles and shells, and weeds and fishes; another tide deposited on top of the sand more carbonized logs, threw them in heaps, and again covered them with marine sediment. The heated ground metamorphosed the charred wood into coal, and if the wood or the ground where it was buried was drenched in a bituminous outpouring, bituminous coal was formed. Wet leaves sometimes survived the forest fires and, swept into the same heaps of logs and sand, left their design on the coal. Thus it is that seams of coal are covered with marine sediment; for that reason also a seam may bifurcate and have marine deposits between its branches.

A support of this my view on the origin of coal I find in a recently published extensive work by Heribert Nilsson, professor emeritus of botany at Lund University.[5] Nilsson presents the results of an inquiry into the botanical and zoological composition of the brown coal (lignite) of Geiseltal in Germany, made by Johannes Weigelt of Halle and his group.[6] Many plants found in Geiseltal lignite are tropical, of species that do not grow even in the subtropics. A long list of tropical families, genera, and species, discerned in Geiseltal coal, was made known (E. Hoffmann; W. Beyn). Algae and fungi on the leaves preserved in the coal are found today on plants in Java, Brazil, and Cameroons (Köck).

Besides the dominating tropical flora in Geiseltal, plants are represented there from almost every part of the globe. The associated insect fauna of Geiseltal coal is found "in present Africa, in East Asia, and in America in various regions, preserved in almost original purity" (Walther and Weigelt). The coal of Geiseltal is rated as belonging to the beginning of the Tertiary time.

As to the reptilian, avian, and mammalian fauna, the coal is a "veritable graveyard." Apes, crocodiles, and marsupials (pouch aniamls) left their remains in this coal. An Indo-Australian bird, an American condor, tropical giant snakes, East Asian salamanders, left their remains there too (O. Kuhn). Some of the animals are of the steppe habitat, and others like crocodiles, came from swamps.

Not only do the origin and the habitats of plants and animals offer a very paradoxical picture, but so also does their state of preservation. Chlorophyll is preserved in the leaves found in the brown coal (Weigelt and Noack). The leaves must have been

[5] H. Nilsson, *Synthetische Artbildung*, 2 vols. (1953), Chaps. VII-VIII.
[6] The papers of Weigelt and his collaborators were published in *Nova Acta Leopoldina*, 1934–41.

rather quickly excluded from contact with air and light, or rapidly entombed: these were neither leaves falling off the plants in the autumn nor leaves exposed to the action of light and atmosphere after being torn off by a storm. Entire strata of leaves from all parts of the world, counted by the billions, though torn to shreds but with their fine fibres (nervature) intact, in many cases still green, are found the Geiseltal lignite.

It is not different with the animals. If exposed after death for any length of time to natural conditions, the structure of animal tissues loses its fineness; the muscles and the epidermis (skin) of the animals of the brown coal of Geiseltal were found to have retained their fine structure (Voigt). Also the colours of the insects are preserved in their original splendour. The very process of fossilization with silica invading the tissues must have occurred "fast blitzschnell"—almost instantaneously, in Nilsson's opinion. While the membranes and the colours of the insects are preserved so well, it is difficult to find a complete insect: mostly only torn parts are found (Voigt).

Nilsson is convinced that the animals and plants found in Geiseltal coal were carried there by onrushing water from all parts of the world, but mainly from the coasts of the equatorial belt of the Pacific and Indian oceans—from Madagascar, Indonesia, Australia, and the west coast of the Americas. One thing is, however, evident: coal originated in cataclysmic circumstances.

Chapter XIV

EXTINCTION

Fossils

Millions of buffaloes have died natural deaths on the prairies of the West in the more than four hundred years since the discovery of America; their flesh has been eaten by scavengers or putrefied and disintegrated; their bones and teeth resisted for a while the decaying process, but finally weathered and crumbled to powder. No bones of these dead buffaloes became fossils in sedimentary rocks, and scarcely any are found in a state of preservation.

The evolutionary theory of the formation of fossils makes certain conditions obligatory: Sedimentary rock is formed in a slow process on the bottom of the sea, and the bones of animals buried in the sediment become fossilized. Land animals wade in the shallow waters of the sea or lakes, die when wading, and their bodies are covered with sediment. The sediment must quickly cover the animals, and this is most possible when the ground subsides. Therefore Darwin postulated such subsidence of the sea bottom as a condition for the formation of fossils. On the other hand, the subsidence or emergence of the ground in the theory of uniformity or evolution is a very slow process, longer by far than the time necessary for a cadaver to disintegrate in water.

The giant reptiles are supposed to have lived as amphibians— on land and in the shallow sea—because of the numerous fossil remains in sedimentary rock. However, no signs of adaptation for aquatic life are found in their skeletons. The bodies were so heavy, it is assumed, that they looked for an opportunity to wade or swim—though it would seem that if they had difficulty in carrying their bodies on land they must have experienced still more difficulty in dragging themselves out of the muddy ground of the shallow water on the beaches. Birds too are supposed to have died while wading and been buried.

When a fish dies its body floats on the surface or sinks to the bottom and is devoured rather quickly, actually in a matter of hours, by other fish. However, the fossil fish found in sedimentary rock is very often preserved with all its bones intact. Entire shoals of fish over large areas, numbering billions of specimens, are found in a state of agony, but with no mark of a scavenger's attack.

The explanation of the origin of fossils by the theory of uniformity and evolution contradicts the fundamental principle of these theories: Nothing took place in the past that does not take place in the present. Today no fossils are formed.

Petrified bones of reptiles, birds, and mammals are often found in large unbroken areas; and since it is quite difficult to describe such areas as wading places, another explanation of the origin of fossils is sometimes offered; the animals were drowned and buried in inundations of large rivers. This explanation seems for certain cases generally closer to the truth than the wading theory; however, the size of the continental areas covered by flood imply catastrophic events on a large scale, and such events, far beyond what is observed on seasonally overflowing rivers today, again contradict the principle of uniformity.

Finally, the very process of sediment formation is not without its problem. Sediment building is supposed to go on permanently in the sea, the building material being the detritus carried by the rivers or broken by the billows from the rocks on the coast and, mainly, the ooze, or calcareous skeletons of myriads of minute living beings, which are abundant in the sea and find their graves on the bottom. The thickness of the sediment on the bottom of the ocean is supposed to give a timetable for the age of the ocean; but, contrary to expectation, in some places on the bottom of the ocean core samples have detected almost no sedimentary rock, indicating that the bottom of the ocean was formed in those places only recently; and in other places, even on land, the sedimentary rock is enormously thick, sometimes tens of thousands of feet. If one and the same process continually and equally deposits the calcareous ooze and detritus on the sea bottom, the inequalities in sedimentary bedrock are as little explained as the formation of fossils.

Both these phenomena are explainable by cataclysmic events in the past. The floor of the ocean was lifted in some places and dropped in others, the sediment was violently shifted, the content of the ocean depths was spilled onto the land, land animals were engulfed and buried by enormous tides carrying debris, in many places avalanches of sand and volcanic dust entombed the aquatic life, fish skeletons remaining in poses of death, undevoured and undecayed.

Footprints

In numerous places and in various formations are found footprints of animals of prehistoric times. Those of dinosaurs and other animals are clearly impressed in rock. The accepted explanation is that these animals walked on muddy ground, and their imprints were preserved as the ground became hard and stony.

This explanation cannot stand up against critical examination. On muddy ground one may find impressions of the hoofs of cattle or horses. But the very next rain will smudge these impressions, and after a short while they will be there no more.

If we do not find the hoofprints of cattle that passed along a path the season before, how is it that the toe imprints of animals of prediluvial times remain intact in the mud on which they walked?

The imprints must have been made like impressions in soft sealing wax that hardens before they are blurred or obliterated. The ground must have been soft when the animal ran upon it, and then it quickly hardened before changes could take place. Sometimes we see imprints of animals that chanced to walk over freshly laid concrete. While the substance was soft, a dog or a bird or a large insect might have walked on it and left impressions recognizable when it hardened. Also, heated sand, turning into a viscous substance on its way to becoming hardened glass, could receive and preserve imprints. The vestiges could also remain in muddy, unheated ground that was soon covered by lava which filled in the imprints and later disintegrated on being weathered away. In historical times—in the volcanic destruction of Pompeii and Herculaneum—lava and volcanic ashes filled the wheel tracks in the streets of these cities and thus preserved them to our day. In the eruption of Kilauea in Hawaii in 1790, when many people lost their lives, and with them a brigade of the Hawaiian army, the footprints of trapped humans and animals were retained in the hardened volcanic ash.[1]

Wherever footprints in the ground dating from historical or prehistoric times are found, we may assume that most probably a catastrophe took place when these vestiges were left or very shortly thereafter. If a catastrophe was in progress or was threatening, the animals must have been in terror and flight. The footprints actually show that the animals in most cases were fleeing, not wading or loitering about; sometimes the configuration of the impressions indicates that an animal was indecisive, probably trapped by some peril closing in from all sides.

The animals that were in flight for their lives may have succumbed a few moments later, crushed or burned in the disaster. The ground was swept by driven sand and ashes or covered by lava or asphalt, or cement, or fluid silicon, then possibly covered by floods, and the imprints in the heated soil that was baked to stone have survived to the present day. So it is that we do not find tracks of animals that peacefully walked one hundred or three hundred years ago, but we do find traces and vestiges of animals that walked and ran many thousands of years ago.

The Caverns

It has been observed that when in a great panic carnivorous animals and the animals that usually are their prey flee together without falling upon each other or being afraid of each other. So when forests burn, horses and wolves, gazelles and hyenas flee along the same paths, all gripped by the same terror, paying no attention to one another. When prairies burn or jungles are

[1] W. M. Agar, R. F. Flint, and C. R. Longwell, *Geology from Original Sources* (1929), Plate XXVIIIB.

enveloped in flame, wild beasts and tame creatures in mixed herds stampede to save their lives. In earthquake or in flood, animals lose their mutual animosity in a common fear. It has also been observed that in earthquakes and other calamities wild animals will come to the abodes of men. In their great migration animals behave differently than when they travel singly or in small herds; so lemming, which scurry away from a man at the sound of his footsteps, will overrun house, town, and river when migrating in large bands, perishing in great numbers but going forward in a huge wave.

In great natural catastrophes animals seek cover from terrifying phenomena—floods, falling meteorites, burning forests, and frightening portents in the sky. Caverns are the places of refuge most sought. An instinct in animals impels them to escape into a den, a hole in the ground, and large animals run for caverns. They may remember such places in the hour of catastrophe, and one may follow another. Of course, many animals never reach the shelter of a cave, but some of them do. And when, in the detritus on the floor of a cave, bones are found of animals that usually would not associate, and the bones are mixed together, and those of the prey animals are not crushed by the teeth of the carnivores, then it is almost certain that these animals tried to save themselves, unafraid of one another, in this cave in the face of approaching catastrophe.

It is possible that some of the animals in the shelter survived the catastrophe, and then their wild instincts must have returned; but in many cases all of them succumbed, overwhelmed by gases, smoke, eddying currents on the surface of the earth, and tides that buried them under sediment.

In numerous places of the world the bone content of caves indicates that they served as hide-outs in times of supreme danger. Lions and tigers, wolves and hyenas, gazelles and hares shared the refuge and there found their common grave. But not all places where such assemblages of bones are discovered were sought for refuge. In many cases the animals were swept from large areas by a tidal wave and thrown against rocks, and the water rushing through the fissures left behind the animals with all their bones broken within their torn bodies. From as far as China, to England and France and the islands of the

Mediterranean, examples of fissures with bones, splintered and mingled together, have been presented in this book.

Not only fissures in the rocks but caverns in the hills may have been filled with bones, though the caverns might not originally have been sought for shelter. An irrupting sea or great lake, lifted from its bed and carrying its own detritus and land debris, swept heterogeneous herds of animals and carried them to the farthest reaches and threw over them hills of gravel, rock, and earth. Cumberland cave, described on an earlier page, is one of many examples.

If the bones are found rolled, they were most probably carried from afar, and were from animals that had died long before; if the bones are more or less intact, the chances are that the place was a shelter that failed; and if the bones are splintered, it is highly probable that the animals were smashed by a great force against rocks or resisting ground.

Extinction

Many forms of life, many species and genera of animals that lived on this planet in a recent geological period, in the age of man, have utterly disappeared without leaving a single survivor. Mammals walked in fields and forests, propagated and multiplied, and then without a sign of degeneration vanished.

"A considerable group have become extinct virtually within the last few thousand years. . . . The large mammals that died out [in America] include all the camels, all the horses, all the ground sloths, two genera of musk-oxen, peccaries, certain antelopes, a giant bison with a horn spread of six feet, a giant beaverlike animal, a stag-moose, and several kinds of cats, some of which were of lion size."[1] Also the Imperial elephant and the Columbian mammoth, animals larger than the African elephant and common all over North America, disappeared. The mastodon that inhabited the forests and ranged from Alaska to the Atlantic coast and Mexico, and the woolly mammoth that roamed in a broad area adjacent to the ice sheets, likewise persisted until a few thousand years ago.[2]

[1] Flint, *Glacial Geology and the Pleistocene Epoch*, p. 523.
[2] L. H. Johnson, *Scientific Monthly*, October, 1952.

The dire wolf, the sabre-toothed tiger, the short-faced bear, the small horse (*Equus tau*) disappeared, and are no longer found either in the Old or in the New World. Many birds, too, became extinct.

These species are believed to have been destroyed "to the last specimen" in the closing Ice Age. Animals, strong and vigorous, suddenly died out without leaving a survivor. The end came, not in the course of the struggle for existence—with the survival of the fittest. Fit and unfit, and mostly fit, old and young with sharp teeth, with strong muscles, with fleet legs, with plenty of food around, all perished.

These facts, as I have already quoted, drive "the biologist to despair as he surveys the extinction of so many species and genera in the closing Pleistocene [Ice Age]."[3]

In the woolly mammoth the genus of elephants achieved its evolutionary perfection; as was already shown by Falconer and known to Darwin, the teeth of the mammoth were superior to those of modern elephants; and in many other respects their adaptation was perfect. The theory of evolution had in the mammoth one of the best examples of a species evolving in the struggle for survival by adaptation. Stone Age man made drawings of it; possibly he even domesticated some of them. In the Neolithic (Stone Age) town of Predmost in Moravia bones of eight hundred to one thousand mammoths were found; their shoulder blades were used in building tombs. On the vast plains of northern Siberia they roamed in herds. They succumbed there as if in one cold night that fell over the land and knew no recess thereafter. They did not die from starvation—their food was found in their stomachs and also between their teeth. The best-preserved body of a mammoth—with even its eyeballs intact—was found in Beresovka, Siberia, eight hundred miles west of Bering Strait. "A fractured hip and fore limb, a great mass of clotted blood in the chest, and unswallowed grass between the clenched teeth, all point to the violence and suddenness of its passing."[4] Did it fall into a pit or was it tossed by hurricanes and floods? It appears that it was "some sudden and unexpected cataclysm,"[5] for the mammoths, together with

[3] Eiseley, *American Anthropologist*, XLVIII (1946), 54.
[4] R. S. Lull, *Organic Evolution* (revised ed.; 1929), p. 376.
[5] Kunz, *Ivory and the Elephant*, p. 236.

rhinoceroses, bison, and others whose bones and teeth make the main substance of the New Siberian Islands, fill the bottom of the Arctic Ocean above Siberia, and lie in the frozen earth of the Siberian tundras. At about the same time the mammoth also perished in Europe and in America.

The mastodon, too, was exterminated at the dawn of the present era. There was no scarcity of their food—it consisted of herbs, leaves, and bark, as is known from the undigested food found within their skeletons. They lived in all parts of the Americas. Over two hundred skeletons were unearthed in New York State. It is not known what brought this widespread group to an end.

Fossil bones of horses indicate that this was a very common animal in the New World in the Ice Age. But when the soldiers of Cortes, arriving at the shores of America, rode their horses which they had brought from the Old World, the natives thought that gods had come to their country. They had never seen a horse.

Of the horses the Spaniards brought to America some went astray, became wild, and filled the prairies, travelling in herds; the land and its vegetation and its climate proved to be exceedingly well suited for the propagation of this animal.

In many parts of the Americas fossil hunters found fossilized bones of horses in great numbers, often imbedded in rock or in lava, which do not differ in shape from the bones of the present-day horse. Why did the horse become extinct with the end of the Ice Age if the climate became so favourable?

In earlier ages there were in America different-looking horses, with three-toed feet, also very small horses the size of cats. However, the one that looked exactly like a modern horse inhabited America and there became extinct only several thousand years before Cortes brought the European horse to the shores of the New World.

Was not the American horse wiped out by man? In our time the American bison (buffalo) was almost destroyed by man, but he used horses to pursue them and firearms to kill them.

C. O. Sauer has advanced the theory (1944) that the terminal Ice Age fauna was destroyed by man, by hunters making fire-drives in pursuit of game. However, Stone Age hunters burning down forests would not have been able to destroy completely many species of animal, leaving not one of the kind from one coast to another and from Alaska to Tierra del Fuego.

F. Rainey, now of the University of Pennsylvania, has observed that "in certain regions of Alaska the bones of these extinct animals lie so thickly scattered that there can be no question of human handiwork involved. Though man was on the scene of the final perishing, his was not, then, the appetite nor the capacity for such giant slaughter."[6] And because of the wholesale and rapid extermination of fauna, "it seems impossible to attribute the phenomenon to the unaided efforts of man."[7] "Even with the known destructiveness of man, however, it is difficult to visualize how these early hunters, armed with puny flint-tipped spears, could have destroyed enough animals to cause complete extinction. But whatever the actual cause or causes may have been, there is no doubt that the end of the ice masses also saw the end of the exotic animals of the same period. . . . The ice cliffs in the background have shrivelled and gone. The trumpeting herds of mammoths and the pounding hooves of the other animals are no more."[8]

L. C. Eiseley of the University of Kansas wrote: "We are not dealing with a single, isolated relict species but with a considerable variety of Pleistocene [Ice Age] forms, all of which must be accorded, in the light of cultural evidence, an approximately similar time of extinction."[9]

Then could it have been a disease that caused the extinction? Or the change in climate because of the termination of the Ice Age? Professor Eiseley finds that epidemic disease or climatic events attendant on the glacial retreat "are sufficient to explain an enormous reduction in the number of a particular species, but are yet inadequate to illuminate the reason for the inability of the species to rebound, in a few years, from its decimated condition."[10] Besides, no known diseasee would attack so many species and genera. And as for the climatic factor, if glacial conditions are the cause, then, according to G. E. Pilgrim, "at approximately the same time we witness a similar extinction of the mammal faunas of Africa and Asia, though in their

[6] Quoted by Eiseley, *American Antiquity*, Vol. VIII, No. 3 (1943), p. 214.
[7] Ibid., p. 212.   [8] Hibben, *Treasure in the Dust*, pp. 58–59.
[9] Eiseley, *American Antiquity*, Vol. VIII, No. 3 (1943), p. 215.
[10] Eiseley, *American Anthropologist*, XLVIII (1946), 54.

case this may not have been caused by glacial conditions."[11]

But even a sudden climatic catastrophe all over the world could hardly have been adequate by itself to account for an extermination so wide and, for many species, so complete. "Climatic change alone is not enough to explain the extinction of the marvellous Pleistocene fauna. There have been other suggestions, such as clouds of volcanic gases which destroyed whole herds of mammals. . . ."[12] Of what dimensions must these clouds have been? They must have covered almost the entire terrestrial globe. But all the volcanoes of the earth, erupting together, would not be sufficient to destroy so many species and genera. Many agents of destruction must have united their forces with the sudden revolution of the climate to wipe out a major part of the animal population of the earth with many genera and species leaving no survivors.

The extermination of great numbers of animals of every species, and of many species in their entirety, was the effect of recurrent global catastrophes. Of some species every animal was exterminated in one part of the world, but a number of animals succeeded in surviving in another part of the world; so the horses and camels of the Americas were destroyed without a survivor, yet in Eurasia, though decimated, they were not exterminated. But many species were completely extinguished, in the Old World as well as in the New—mammoths and mastodons and others. They expired not because of lack of food or inadequate organic evolution, inferior build or lack of adaptation. Plentiful food and superb bodies and fine adaptation and solid procreation, but no survival of the fit. They died as if a wind had snuffed life out of all of them, leaving their cadavers, with no sign of degeneration, in asphalt pits, in bogs, in sediment, in caverns. Some of the decimated species probably endured for a while, possibly for several centuries, being represented by a few specimens of their kind; but in changed surroundings, amid climatic vicissitudes, with pastures withered, with plants that had served as food or animals that had served as prey gone, these few followed the rest in a losing battle for existence,

[11] G. E. Pilgrim, "The Lowest Limit of the Pleistocene in Europe and Asia," *Geological Magazine* (London), Vol. LXXXI, No. 1, p. 28.
[12] Hibben, *Treasure in the Dust*, p. 59.

surrendering at last in the struggle for survival of a species.

Burning forests, trespassing seas, erupting volcanoes, submerging lands took the major toll; impoverished fields and burned-down forests did not offer favourable conditions for frightened and solitary survivors, and claimed their own share in the work of extinction.

Chapter XV

CATACLYSMIC EVOLUTION

Catastrophism and Evolution

The theory of evolution dates back to the age of classic Greece, one of its proponents having been Anaximander, and from time to time philosophers have offered the evolutionary explanation of the origin of the multiple forms of life on earth, as opposed to the theory of special creation or the permanency of living forms from the day of Creation. Lamarck (1744–1829) thought that acquired characteristics were transmissible by heredity and thus might lead to the appearance of new forms of life. In 1840, the year that Agassiz's Ice Age theory was published, an anonymously printed work, *Vestiges of Creation*— written by Robert Chambers—caused a stir that did not subside for years. It was bitterly attacked by every British scientist for teaching that human beings are "the children of apes and the breeders of monsters," in the words of one critic, the president of the Geological Society, Adam Sedgwick. Darwin later acknowledged that the brunt of the attack against his own theory was absorbed by *Vestiges*.

What was new in Darwin's teaching was not the principle of evolution in general but the explanation of its mechanism by natural selection. This was an adaptation to biology of the Malthusian theory about population growing more quickly than the means of existence. Darwin acknowledged his debt to Malthus, whose book he read in 1838. Herbert Spencer and Alfred R. Wallace independently came to the same views as Darwin, and the expression "survival of the fittest" was Spencer's.

Darwin wrote his theory with the point of his pen directed

against the theory of catastrophism. He hardly expected that no opposition would come from the side he attacked, otherwise he would not have expended so many arguments in combating catastrophism and in subscribing so completely to Lyell's theory of uniformity in lifeless nature. As it turned out, most of the attacks against Darwin came from the Church, which could not agree that man had risen from inferior beings. The Church held to the dogmas of creation in six days less than six thousand years ago, and of the primal sin of Adam, to redeem humankind from which, the Son of Man came into this world; also to the view that beasts have no souls and therefore a barrier stands between man and animal.

The emotions of this protracted controversy were spent on the issue: Is there evolution or is there not? More and more scientists subscribed to evolution; religious minds clung to the belief that there had been no change since the creation of the world. Actually the debate was between liberals and conservatives in the matter of science. The radicals did not participate; for catastrophism was dying out with the generation of the founders and classicists of geological science. Cuvier died in 1832; in England, geologists like Buckland of Oxford and Sedgwick of Cambridge, set in their belief in Mosaic tradition, ascribed the ubiquitous vestiges of the catastrophe to the action of the Deluge. But they could not point to a satisfactory physical cause of such catastrophe, and expert estimate made it obvious that, had all the clouds over the earth emptied themselves simultaneously, the earth would not have been covered by even one foot of water.

Then the geological record showed that there had been not one but several deluges. Lyell wrote in a letter: "Conebeare [geologist and Bishop of Bristol] admits three deluges before the Noachian! and Buckland adds God knows how many catastrophes besides, so we have driven them out of the Mosaic record fairly."[1] Sedgwick, according to Lyell, "decided on four or more deluges."[2] In his last address as president of the Geological Society, Sedgwick admitted that his religious beliefs caused him to propagate a philosophic heresy: "I think it right, as one of my last acts before I quit this Chair, thus publicly to read my recantation. We ought, indeed, to have paused before we adopted the diluvian theory, and referred all our old superficial gravel to

[1] *Life, Letters and Journals*, I, 253.　[2] Ibid.

the action of the Mosaic Flood. For of man, and the works of his hands, we have not yet found a single trace among the remnants of a former world entombed in these deposits."[3]

So where were the remains of the sinful population? Cuvier taught that man's remains were never found with those of extinct animals. Lyell also declared in the first edition of his *Principles* that man was created after all the extinct animals passed away; and not until 1858, a year before the publication of Darwin's *Origin of Species*, did the finds in the Brixham cave shatter this belief in the non-coexistence of man and extinct, or "antediluvian," animals.[4] In the year of the *Origin*, the leading English geologists were finally convinced by J. B. de Perthes, a notary of Abbeville in France, who for twenty years found only deaf ears, that human artifacts (worked flint) and extinct animals are met in the same formations, side by side. This opened wide the doors to Darwin's theory. By that time the doubts of the catastrophists, who could not understand why there were signs of more than one deluge and why there should be no human bones left of all the sinful generation that perished in the Flood, had already brought about the abandonment of the theory of catastrophism, a theory that appeared to be in conflict with the Mosaic record.

Thus it happened that the entire controversy for and against Darwinsim failed to respond to the challenge of Darwin, who tried to show that what appeared to be the result of global catastrophes could be explained as the product of slow changes multiplied by time, with no violence intervening. The opposition was concentrated against the idea of evolution and in support of special creation. Insisting that all animals were created in the forms in which they are found in our days, the opponents of evolution waged their battle on geologically indefensible ground.

But why did Darwin oppose the idea of great catastrophes in the past, contrary to his own field observations, and subscribe

[3] C. C. Gillispie, *Genesis and Geology* (1951), pp. 142–43; Sedgwick, "Presidential Address (1831)," *Proceedings of the Geological Society*, I, 313.

[4] As early as 1832, Sir Henry T. de la Beche in his *Geological Manual*, p. 173, claimed the coexistence of man with extinct animals, because caverns closed by "fragments of rock transported from a distance" contain the remains of man and extinct animals; "he existed previous to the catastrophe which overwhelmed him and them."

to the theory of uniformity of geological events in all ages and in the present? For species to evolve as a result of incessant competition and struggle for survival, all the way from the simplest forms to *Homo sapiens* and other advanced organisms, an enormous span of time is required. The teaching of catastrophes appeared to make the story of the world very short: if the Deluge occurred less than five thousand years ago, then, following the book of Genesis, Creation took place less than six thousand years ago. In order to have at the disposal of the evolutionary process the almost unlimited time needed, Darwin accepted Lyell's teaching; and whereas Lyell tried to show that the usual agents—such as rivers carrying sediment—act with comparative speed, Darwin liked to stress their sluggishness.

He wrote: "Therefore a man should examine for himself the great piles of superimposed strata, and watch the rivulets bringing down mud, and the waves wearing away the seacliffs, in order to comprehend something about the duration of past time." The waves of the sea reduce a rock particle by particle, and if a visible change is produced, it requires many thousands of years.

"Nothing impresses the mind with the vast duration of time, according to our ideas of time, more forcibly than the conviction thus gained that subaerial agencies which apparently have so little power, and which seem to work so slowly, have produced great results."[5] Darwin even went so far as to suggest that "he who can read Sir Charles Lyell's grand work on the *Principles of Geology* . . . and yet does not admit how vast have been the past periods of time, may at once close this volume [*Origin of Species*]."[6]

The Geological Record and Changing Forms of Life

His thesis of the origin of species by natural selection Darwin supported by reference to (1) variations in domestic animals, especially when the breeder deliberately develops a certain desirable feature; (2) the anatomical similarity of many related species; and (3) the geological record. However, though breeders have created new races or variations, they have

[5] *The Origin of Species*, Chap. X.   [6] Ibid.

created no new animal species. In the anatomy of living creatures "the distinctness of the specific forms, and their not being blended together by innumerable transitional links, is a very obvious difficulty" (Darwin); and thus the entire weight of proof was placed on the geological record.

This record shows, however, "The Forms of Life Changing Almost Simultaneously throughout the World"—the title of a section in *The Origin of Species*. Darwin wrote: "Scarcely any palaeontological discovery is more striking than the fact that the forms of life change almost simultaneously throughout the world," This appears baffling, because according to his theory "the process of modification must be slow, and will generally affect only a few species at the same time; for the variability of each species is independent of that of all others." Could it not have been a sudden change in physical conditions that altered the forms of life at one and the same time throughout the world? Darwin answers, No. "It is, indeed, quite futile to look to changes of currents, climate, or other physical conditions, as the cause of these great mutations in the forms of life throughout the world, under the most different climates." If the climate or other physical conditions changed in one part of the world, how could this alter forms of life in all the other parts of the world? That a change in physical conditions could have occurred all over the world at one and the same time, Darwin did not even take into consideration. What kind of an answer to his problem, therefore, could Darwin propose?

"Blank intervals of vast duration, as far as the fossils are concerned, occurred. . . . During these long and blank intervals I suppose that the inhabitants of each region underwent a considerable amount of modification and extinction. . . ." Hence the parallelism of changes in fauna and flora in similar strata around the world is not a true time-parallelism. "The order would falsely appear to be strictly parallel."

Darwin then considered "The Absence of Numerous Intermediate Varieties in Any Single Formation," and wrote: "If we confine our attention to any one formation, it becomes much more difficult to understand why we do not therein find closely graduated varieties between the allied species which live at its commencement and at its close." And he found the answer in the conjecture that "although each formation may mark a very long lapse of years, each probably is short compared with the period requisite to change one species into another."

Furthermore, the geological record shows "The Sudden Appearance of Whole Groups of Allied Species" (the title of another section in *The Origin of Species*). "The abrupt manner in which whole groups of species suddenly appear in certain formations, has been urged by several palaeontologists—for instance, by Agassiz, Pictet, and Sedgwick—as a fatal objection to the belief in the transmutation of species. If numerous species, belonging to the same genera or families, have really started into life at once, the fact would be fatal to the theory of evolution through natural selection. For the developmet by this means of a group of forms, all of which are descended from some one progenitor, must have been an extremely slow progress; and the progenitors must have lived long before their modified descendants."

Darwin explained this observation, too, by the incompleteness of the geological record, which, because of the lacunae, gives the appearance of sudden changes.

The geological record of extinction of species is discussed under the heading, "On Extinction." Darwin wrote: "The extinction of species has been involved in the most gratuitous mystery." What took place is "apparently sudden extermination of whole families or orders." According to his theory, "the extinction of a whole group of species is generally a slower process than their production," and yet some groups were exterminated "wonderfully sudden." Here, once more, Darwin thought that the imperfection of the geological record may in some cases simulate the suddenness of the extinction; but he acknowledged in other cases his inability to explain the spontaneity of the extinction of some species. He still wondered, as in the days of his South American travels, why horses had disappeared in pre-Columbian America where they had every favourable condition for propagation; and in a letter to Sir Henry H. Howorth he acknowledged his inability to explain the extinction of the mammoth, a well-adapted animal. But in general the deficiency of the geological record was invoked to explain the apparent spontaneity of extinction as well as the suddenness with which new species seem to have arrived on the scene.

According to the theory of natural selection, chance variations or new characteristics among individuals of a species, if beneficial, are exploited in the struggle for survival and, being inheritable, may by accumulation lead to the origin of a new species. Because of the chance nature of these new characteristics and therefore also of the origin of the new species, Darwin assumed "that not only all the individuals of the same species have migrated from some one area, but that allied species, although now inhabiting the most distant points, have proceeded from a single area—the birthplace of their early progenitors. . . . The belief that a single birthplace is the law seems to me incomparably the safest."

Darwin explained the migration of plants from continent to continent and from mainland to islands by the transportation of seeds in the intestines of birds; the migration of molluscs, by observed instances of small shells clinging to the legs of migrating birds. This method of dispersion does not account for the geographical distribution of larger animals unable to fly or swim across the sea, or traverse climatic zones unsuitable for the species.

Since animals of such species are found in very distant parts of the globe, divided by oceans, Darwin was led to maintain that "during the vast geographical and climatal changes which have supervened since ancient times, almost any amount of migration is possible." This makes necessary the existence of land connections or "land bridges" between islands and mainlands and between all continents. But to these geographical and climatal changes, the Ice Age included, Darwin ascribed "a subordinate" role in shaping the development of the animals; they played an important role only in the migration of the animals.

Where the land is continuous, as in the Americas, Darwin accounted for the fact that identical animals live in higher latitudes of the Southern and Northern hemispheres, though they are absent in temperate and tropical latitudes, by resorting to a theory which assumes that the glacial periods in the Northern and Southern hemispheres were not simultaneous but consecutive. When a glacial period was descending upon the north, animals migrated slowly to the south, toward the equator; when the glacial period ended, and the climate in the subtropics became hot, some animals returned to the north, others remained in the subtropical regions, climbing the cool mountains. When the next glacial age—this time advancing from the south—arrived, the animals on the mountains came down, and when this age also ended, some of them moved to the south, while others again retreated to the mountains. Thus identical animals are found in the cooler regions of both the Northern and Southern hemispheres. (At present this view of consecutive glacial periods in the Northern and Southern hemispheres has hardly any adherents.)

The theory of evolution by natural selection could not do well without the theory of the ice ages. It needed the Ice Age theory to explain the provenience of the same species in the Southern and Northern hemispheres separated by the Torrid Zone; it needed it even more to account for the phenomenon of drift. Erratic blocks could have been explained, with some straining, by the action of icebergs. But drift, or accumulation of clay, boulders, and sand that in many places fills valleys hundreds of feet deep, could not have been brought in by icebergs; and, finally, icebergs, in order to be produced in great numbers, themselves required extended glaciers from which they could break off. Darwinian evolution needed the Ice Age theory in order to supplant the tidal wave theory—which is a catastrophic notion.

Darwin accepted Agassiz's teaching, though not in its original form with a catastrophic beginning of the ice ages. But Agassiz rejected Darwin's theory. The reason for this he saw in the skeletal remains of ancient fish, a field in which he was an authority. In many instances the fish of extinct species were better developed and further advanced in their evolution than later species, the modern included. Among mammals, too, many better-developed species became extinct. But these difficulties in the way of the evolutionary theory were less strongly felt in the heat of the fight against the opponents who insisted on a six-thousand-year-old world and the immutability of species.

Darwin's theory represented progress as compared with the teachings of the Church. The Church assumed a world without change in nature since the Beginning. Darwin introduced the principle of slow but steady change in one direction, from one age to another, from one aeon to another. In comparison with the Church's teaching of immutability, Darwin's theory of slow evolution through natural selection or the survival of the fittest was an advance, though not the ultimate truth.

The story of his experiences is told by his contemporary and adherent, Thomas Huxley. Darwin was "held up to scorn as a 'flighty' person, who endeavours 'to prop up his utterly rotten fabric of guess and speculation,' and whose 'mode of dealing with nature' is reprobated as 'utterly dishonourable to Natural Science'." Thus Huxley quoted from an article by Bishop Wilberforce in the *Quarterly Review* of July 1860. Huxley also wrote in 1887: "On the whole, then, the supporters of Mr. Darwin's views in 1860 were numerically extremely insignificant. There is not the slightest doubt that, if a general council of the Church scientific had been held at that time, we should have been condemned by an overwhelming majority. And there is as little doubt that, if such a council gathered now, the decree would be of an exactly contrary nature."

Darwin's *Origin of Species*, Huxley went on "was badly received by the generation to which it was first addressed, and the outpouring of angry nonsense to which it gave rise is sad to think upon. But the present generation will probably behave just as badly if another Darwin should arise, and inflict upon them that which the generality of mankind most hate—the necessity of revising their convictions. Let them, then, be charitable to us ancients; and if they behave no better than the men of my day to some new benefactor, let them recollect that, after all, our wrath did not come to much, and vented itself chiefly in the bad language of sanctimonious scolds. Let them as speedily perform a strategic right-about-face, and follow the truth wherever it leads. The opponents of the new truth will discover as those of Darwin are doing, that, after all, theories do not alter facts, and that the universe remains unaffected even though texts crumble."[1]

[1] Thomas H. Huxley, "On the Reception of the *Origin of Species*," printed as Chap. XIV of the first volume of *The Life and Letters of Charles Darwin*, ed. by his son Francis Darwin, in the Appleton edition of the *Works* of Charles Darwin.

The Mechanism of Evolution

Natural selection—the Darwinian mechanism of evolution—is simultaneously destructive and constructive. In the struggle for existence it eliminates all the unfit among the members of a species; and it destroys the species that cannot compete with others for the limited resources of livelihood. The winners in this struggle are those individuals that because of some characteristic—or favourable variation—have an edge over other competitors. "Under these circumstances favourable variations would tend to be preserved, and unfavourable ones destroyed. The result of this would be the formation of new species" (Darwin).

As shown on previous pages, the annihilation of many individuals and of entire species in the animal kingdom took place, not only under circumstances of competition, but under catastrophic conditions as well. Entire species with no sign of degeneration suddenly came to their end in paroxysms of nature. Yet extinction of a species through starvation or extermination by enemies also takes place: Moa, the gigantic flightless bird of New Zealand that stood twelve feet high, was destroyed several centuries ago. The whooping cranes of North America were reduced by 1953 to twenty-one individuals. Natural selection cannot account for the wholesale destruction of many genera and species at one time; it may occasionally be the agent exterminating single species. But can natural selection create new species?

The geological record presents evidence that in the past animals lived that do not live any longer; and also that, of the forms living today, many did not exist in the past. Then how did they come into being?

The animal and plant kingdoms are subdivided into phyla, and these into classes, orders, families, genera, and finally species. A species can be recognized this way: the mating of members of two different species generally does not produce offspring, and when it does, such offspring is sterile (horse and ass, and their offspring, the mule). Thus all the human race is but one species; and all races of dogs, so dissimilar in their body structures, are members of one species. There are hundreds of thousands of species in the animal kingdom and also in the plant kingdom.

In the theory of evolution all forms of life evolved by gradual emergence from the same most primitive, one-cell living beings. Chance variations occur in members of every species—no two individuals are entirely identical. These variations are inheritable. As already explained, the favourable variations—those that are helpful in the struggle for existence—may accumulate to such a degree that, according to Darwin, a new species originates, the members of which can have no fruitful progeny with the members of the parental species.

Since the first scientific observations were made, no truly new animal species has been observed to come into being. The year after publication of *The Origin of Species*, Thomas Huxley wrote: "But there is no positive evidence, at present, that any group of animals has, by variation and selective breeding, given rise to another group which was, even in the last degree, infertile with the first."[1] A few years later Darwin wrote in a letter (to Bentham): "The belief in natural selection must at present be grounded entirely on general considerations. . . . When we descend to details . . . we cannot prove that a single species has changed; nor can we prove that the supposed changes are beneficial, which is the groundwork of the theory."[2] And at the end of the century Huxley found himself compelled to make the statement: "I remain of the opinion . . . that until selective breeding is definitely proved to give rise to varieties infertile with one another, the logical foundation of the theory of natural selection is incomplete. We still remain very much in the dark about the causes of variation. . . ."[3]

In selective breeding the breeder creates conditions not found in wild life; and new races or varieties of animals created by selection and isolation revert to their ancestral unselected forms as soon as they are turned free; thus when dogs of various breeds mate they give birth to mongrels which resemble their common ancestors. Despite all their efforts, breeders have not been able to cross the true frontier of a species. Then how could

[1] Thomas H. Huxley, "The Origin of Species" (1860), reprinted in his *Darwiniana, Collective Essays* (1893), II, 74.
[2] Darwin, *Life and Letters*, ed. Francis Darwin, II, 210.
[3] Huxley, *Darwiniana, Collective Essays* (1893), II, Preface.

a new species originate in chance variations and through cross-breeding in wild life? And how could so many new species be produced that they number, together with the extinct, in the millions? And how could a human being, so complicated, evolve, not just from common ancestors with the primates (apes), but from common ancestors with winged insects and crawling worms? The evolutionists drew more checks on time.

Then, too, the chance character of variations, when they first appear in an individual, makes the envisaged progress especially difficult. Darwin professed ignorance as to the cause of these variations or new characteristics appearing in individuals; and it was generally understood that chance variations, in the vast majority of cases, must be in the nature of defects: in a complicated and balanced organism a chance variation would probably be a hindrance, not a benefit. Then by what rare accidents could ever more perfected species have originated?

Various theories have been offered—one of them being *évolution créatrice* by Henri Bergson—that assume the existence of a guiding principle in evolution, which replaces the chance and accident in variations; these theories are often united under the name *orthogenesis*, the best known of such ideas. The adherents of orthogenesis claim the existence of a plan and a goal. But since, in such a theory, Providence enters into action, and to make nature independent of it was a major objective of the theory of evolution as opposed to the teaching of special creation, after some deliberation orthogenesis, or creative evolution, met largely with rejection. The orthogeneticists could argue that many traits, when they first appeared, must have been entirely useless, yet not senseless if they were destined to become useful after many generations. Then why should these traits have gone on developing from age to age, finally to become an asset to the species, unless orthogenesis was in action; why should the pocket of the kangaroo have increased in size through many generations until it could be used for carrying baby kangaroos?

The obvious difficulty in explaining the evolutionary process by chance variations brought about the revival of Lamarckism. In 1809, the year Darwin was born, Lamarck had published his *Philosophie zoologique*, in which he offered a theory of evolution through the appearance of new traits and faculties in response

215

to usage; usage in response to need; and need as the consequence of changes in physical surroundings. These new acquired traits, he assumed, were inheritable. Lamarck also taught uniformity, and thus he was an opponent of his contemporary, Cuvier, who taught catastrophism. Charles Darwin, generous to Alfred R. Wallace, whom he declared to be an independent discoverer of the theory of natural selection, never agreed, despite the admonitions of Lyell and Huxley, to acknowledge his debt to Lamarck; in a letter to Lyell he referred to Lamarck's book as "absurd" and "rubbish," and also as a "wretched book."[4] However, Darwin offered the theory of *pangenesis*, according to which every cell in the body of an animal or plant sends a gemmule, an invisible image of the parent cell, to the germ cells. In this way Darwin intended to interpret heredity. Thus he went even farther than Lamarck in making the cells of the body the carriers of heredity, which amounts to hereditary transmission of acquired traits. The theory of pangenesis is definitely rejected by everyone.

In the battle that went on among the representatives of different schools in evolution, the neo-Darwinists, led by August Weismann, attacked the neo-Lamarckists; and by cutting off the tails of mice in succeeding generations, Weismann could show that acquired traits are not inheritable. Actually, he did not prove that much: the loss of tails by cutting is not a habit or trait acquired through usage or need. It was Weismann who really disproved Darwin's pangenesis theory, not Lamarck, but he properly stressed that the carriers of heriditary traits are in the germ plasma, or in spermatozoa and ova; the soma, or the body, is created in each successive generation by the germ plasma, and only changes in the plasma are inheritable. The chance variations of Darwin are such changes in the germ plasma and are therefore inherited; the response of the body to external agents would not create inheritable traits and therefore must be of no value in evolution.

On evolution as a geological fact all agreed, but on the mechanism of evolution the disagreement has been fundamental. The majority of evolutionists have rejected the idea that acquired characteristics are inheritable; but Lamarck's ideas

[4] Darwin, *Life and Letters*, II, 199; L.T. More, *The Dogma of Evolution* (1925), p. 172.

found followers in the East, in Michurin, who experimented on plants, and for a time in Pavlov, who experimented on animals, and not long ago in the dominant school of thought in Russia.

The neo-Darwinists deny that physical surroundings can give rise to new species; they may bring about changes in an organism, but the acquired characteristics are not inheritable. Can, then, natural selection or competition with other animals create new species? The classic example of a giraffe with the longest neck surviving when leaves are left only high on the trees does not prove that giraffes with longer necks would become a separate species. And, in any event, under the described conditions no new race would evolve: the female giraffe, which are smaller in stature, would die out before the male competitors, and there would be no progeny; but should there be progeny, the young giraffes would probably die because they would be unable to reach the leaves.

The position of Darwinists would be much stronger if a new animal species would appear, even if only in controlled breeding. Darwin claimed that the process of the appearance of new species is very slow, but he also maintained that the process of extinction of a species is even slower.[5] Nevertheless, some species of animals have expired before the eyes of the naturalists, but no new one has appeared. The theory of natural selection, even the very fact of the evolvement of one species from another, needed proof. Some scientists went so far as to say that possibly the entire development plan has already reached its permanent stage, and the geological records tell only of the road to that stage, evolution no longer taking place.

One part of the Darwinian theory of selection has been generally abandoned: it is the idea of sexual selection as a factor in evolution. In natural selection the competition is for the means for existence. In sexual selection—a theory developed in *The Descent of Man* (1871)—the competition is among the males for acceptance by a female. Darwin thought to explain the origin of various secondary sexual characteristics, such as ornamentation and colour of feathers in birds, by saying that they were the results of gradual selection, through many generations, of traits attractive in the eyes of the female. But it was

[5] *The Origin of Species*, Chapter XI.

217

shown that when the colourful wings of male butterflies were cut off and in their stead female wings, often without the characteristic colouring, were glued to the body of the male butterfly, the female did not object to the approach of the male. She failed to discriminate against male butterflies with no wings at all. Also it was observed that some male fish fertilize the fish eggs, having all the male colouring characteristics of such season, but without the female fish being present or aware of the act of fertilization. The theory of sexual selection to a certain degree had the same fate as the theory of gemmules. But the theory of natural selection would not yield its position unless a better explanation of the evolutionary mechanism could be given.

Mutations and New Species

The first ray of light came at the turn of the century, when Hugo De Vries, a Dutch botanist, observed spontaneous mutations in the evening primrose. The plant, without a recognizable cause, would show new characteristics unobserved in its ancestors. Although De Vries claimed that these mutations amount to what may be called "little species," they have not caused the primrose to pass beyond the frontier of its species. However, it was demonstrated that variations within a species do appear in a spontaneous manner, and rather suddenly, and not, as Darwin thought, by minute progressions from generation to generation. Huxley was correct in urging Darwin not to adhere so dogmatically to his belief that nature does not make jumps—*natura non facit saltum*.[1] De Vries showed that variations are in the nature of jumps, and from this he developed the mutation theory of evolution.

De Vries, while working on his theory, was as yet unaware of Gregor Mendel's investigations in genetics, already published as a paper in 1865, only six years after *The Origin of Species*. Mendel's work, unknown to Darwin and his followers in the nineteenth century, was rediscovered by De Vries and independently by E. Tschermak and K. Correns in 1900, the same year that De Vries wrote down his theory of mutations. By carefully observing crossings between varieties of the garden pea and

[1] Darwin, *Life and Letters*, II, 27.

counting the strains through consecutive generations and the transmission of single traits, Mendel established the fundamental laws of genetics or inheritance of somatic characteristics. The entire work on evolution since the beginning of this century is based on genetics and Mendel's laws. Ironically, Mendel was an Augustine monk and made his basic contribution at a time when the war between science and the Church was raging, following the publication of Darwin's main work. The spontaneous variations in mutants can be followed through as hereditary factors in successive generations of offspring. The genes in the germ plasma are the carriers of the traits, and a variation (mutation) in a gene would cause a variation (mutation) in the offspring. But, generally, only single variations appear at a time; they may lead to new races, not to new species.

Spontaneous mutations are far too few and insufficient in magnitude to bring about the appearance of new species and to explain how the world of animals came into existence. Despite all spontaneous variations no new species of mammals are known to have been created since the close of the Ice Age. In 1907, V. L. Kellogg of Stanford University came to the following conclusion:

"The fair truth is that the Darwinian selection theories, considered with regard to their claimed capacity to be an independently sufficient mechanical explanation of descent, stand today seriously discredited in the biological world. On the other hand, it is also fair truth to say that no replacing hypothesis or theory of species forming has been offered by the opponents of selection which has met with any general or even considerable acceptance by naturalists. Mutations seem to be too few and far between; for orthogenesis we can discover no satisfactory mechanism; and the same is true for the Lamarckian theories of modification by the cumulation, through inheritance, of acquired or ontogenic characters."[2]

Kellogg also observed that one group of scientists "denies in toto any effectiveness or capacity for series forming on the part of natural selection, while the other group, a larger . . . sees in natural selection an evolutionary factor capable of initiating nothing, dependent wholly for any effectiveness on some primary factor or factors controlling the origin and direction of

[2] V. L. Kellogg, *Darwinism Today* (1907), p. 5.

variation, but capable of extinguishing all unadapted, unfit lines of development. . . . For my part," Kellogg concluded, "it seems better to go back to the old and safe Ignoramus standpoint." Thus the entire problem was shunted back to the place it occupied before *The Origin of Species*.

Evolution is the principle. Darwin's contribution to the principle is natural selection as the mechanism of evolution. If natural selection, sharing the fate of sexual selection, is not the mechanism of the origin of species, Darwin's contribution is reduced to very little—only to the role of natural selection in weeding out the unfit.

H. Fairfield Osborn, a leading American evolutionist, wrote: "In contrast to the unity of opinion on the *law* of evolution is the wide diversity of opinion on the *causes* of evolution. In fact, the causes of the evolution of life are as mysterious as the law of evolution is certain,"[3] And agan: "It may be said that Darwin's law of selection as a natural explanation of the origin of *all* fitness in form and function has also lost its prestige at the present time, and all of Darwinsim which now meets with universal acceptance is the law of the survival of the fittest, a limited application of Darwin's great ideas as expressed by Herbert Spencer."[4]

These were not the opinions of single evolutionists, but generally held views. William Bateson, a leading English evolutionist, in his address before the American Association for the Advancement of Science in 1921, said:

"When students of other sciences ask us what is now currently believed about the origin of species we have no clear answer to give. Faith has given place to agnosticism. . . . Variation of many kinds, often considerable, we daily witness, but no origin of species. . . . I have put before you very frankly the considerations which have made us agnostic as to the actual mode and processes of evolution."[5]

L. T. More, in a series of guest lectures delivered at Princeton University, asked:

"If natural selection is a force which can destroy but cannot

[3] Henry Fairfield Osborn, *The Origin and Evolution of Life* (1917), p. ix.
[4] Ibid., p. xv.
[5] William Bateson, "Evolutionary Faith and Modern Doubts," *Science*, LV, 55.

create species and if the reasons for this destruction are un-known, of what value is the theory to mankind? . . . The col-lapse of the theory of natural selection leaves the philosophy of mechanistic materialism in a sorry plight."[6]

On De Vries's theory of evolution by mutations More said: "The idea is destructive to scientific theory, as it really does away with the whole idea of continuity which should be the basis of an evolution. . . . The thought at once occurs that each of the surprising breaks in the paleontological record, such a one as separates the reptile from the feathered bird, may have been taken at a single leap during an overstimulated period of nature."[7]

De Vries made observations of spontaneous mutations in plants; a decade later T. H. Morgan found spontaneous muta-tions in *Drosophila melanogaster*, the vinegar fly, including various colourings of the eyes and various lengths of wings, and many other changes in progeny not present in any of the ancestors. H. J. Muller, by subjecting the vinegar fly to the action of X-rays, increased the frequency of mutations one hundred and fifty times. It was also found that some chemicals and tempera-tures close to the limits that the insect organism can endure may act as mutation-provoking agents.

Muller concluded that spontaneous mutations are "usually due to an accidental individual molecular or submolecular collision, occurring in the course of thermal agitation," and this is indicated "by the amount of rise in the frequency of mutations that is observed when the temperature is raised, so long as tem-peratures normal to the organism are not transgressed. Since chemical changes similar to but more extreme than those of thermal agitation may also be produced by X-rays and other high-energy radiation and by ultra-violet, it is not surprising that mutations like the so-called 'spontaneous' ones can be induced in great abundance by these means, and that the num-ber of these mutations is, in general, proportional to the number of physical 'hits' caused by the radiation."[8]

The origin of mutations in the evening primrose, observed by De Vries, like every other spontaneous mutation, must be

[6] More, *The Dogma of Evolution* (1925), p. 240. [7] Ibid., p. 214.
[8] Muller, "The Works of the Genes," in H. J. Muller, C. G. Little, and L. H. Snyder, *Genetics, Medicine and Man* (1947), p. 27.

221

ascribed to one of those irritants acting directly on the genes. It could have been the result of hits by cosmic rays; only it must be shown why the evening primrose is more susceptible to such an agent than most other plants.

The practical absence of X-rays in surrounding nature caused this powerful agent of mutations in laboratories to be regarded as not operative in spontaneous mutations and therefore also not in the process of evolution. Muller stressed this point. However, an X-ray component is present in radium radiation. At the beginning of the present century it was noticed that tadpoles or embryonic frogs in the presence of a tube containing radium give rise to various freaks.[9] Radioactivity and cosmic radiations are agents present in nature, one of terrestrial, the other of extraterrestrial, origin.

If, as the experiments with the vinegar fly demonstrated, a mutation of some gene can produce a wingless fly, many mutations simultaneously or in quick succession would be quite able to transform an animal or plant into a new species. In the bomb craters of London new plants, not previously known on the British Isles, and possibly not known anywhere, were seen to sprout. "Rare plants, unknown to modern British botany, were discovered in the bomb craters and ruins of London in 1943."[10] It appears that the thermal action of bomb explosions was the cause of multiple metamorphoses in the genes of seeds and pollens. If this is so, then the statement made earlier that no new species has been observed in the process of making its first appearance must be retracted.

It must be retracted anyway, so far as the plant (not animal) kingdom is concerned, in view of the claims made by a certain school of plant geneticists that now and then some plants may produce an abnormal offspring with a double number of chromosomes and, further, that, though hybrids in plants, as in animals, generally have no offspring, hybrids from the double-chromosome parents may occasionally produce a true new species; it can reproduce itself indefinitely, but it cannot reproduce by crossing with the original species, or if it does the offspring resulting from the crossbreeding is sterile. An alkaloid

[9] R. H. Bradbury, "Radium and Radioactivity in General," *Journal of the Franklin Institute*, Vol. CLIX, No. 3 (1905).

[10] "Botany," *Britannica Book of the Year, 1944*, p. 117.

(colchicine) from the roots of the autumn crocus, when applied to cells in the process of division, helps to produce cells with twice the normal number of chromosomes. Thus a fertile cross between the radish and the cabbage was achieved, and the proponents of "cataclysmic evolution" claim that chance appearance of double-chromosome plants was responsible in the past for the origin of cultivated wheat, oats, sugar, cane, cotton, and tobacco, and will permit production in the laboratory of a grain that would combine the desirable qualities of both wheat and rye. What causes a plant to produce spontaneously an offspring with a double number of chromosomes is not yet sufficiently known; and most probably, again, thermal, chemical, or radioactive agents are involved.

Cataclysmic Evolution

When, therefore, the earth, covered with mud from the recent flood, became heated up by the hot and genial rays of the sun, she brought forth innumerable forms of life, in part of ancient shapes, and in part creatures new and strange.

—OVID, *Metamorphoses* (trans. F. J. Miller)

An enormous expansion of radioactivity in bygone ages was postulated by various theorists as an explanation of great oscillations in climate in the past; the thermal effect of widespread radioactivity is likewise claimed as a motive force by the author of the modern version of the theory of drifting continents (Du Toit). It appears to me that if such radioactivity really occurred its mutation effect could not have failed to take place too.

Cosmic rays or charges, hitting nitrogen atoms in the atmosphere, transform this element into radiocarbon. These charges, arriving from outside the earth, are very strong per particle, averaging several billions of electron volts and sometimes carrying a potential of a hundred billion electron volts. As comparatively few such rays or charges hit our atmosphere, their general effect is not spectacular. But it is conceivable that, where a cosmic ray or charge hits a gene of germ plasma, a biological mutation takes place, comparable to the physical

223

transmutation of the elements. After all, the genes, like any proteins, are biochemical compounds composed of carbon, nitrogen, and a few other elements. Should a somatic chromosome be hit by a powerful charge, it might at worst cause disorganized growth and be the origin of a neoplasma; but if the genes of the germ plasma should be the target of a collision with a cosmic ray or secondary radiation, a mutation in the progeny might ensue; and should many such hits occur, the origin of a new species, most probably incapable of individual or genetic life, but in some cases capable, could be expected. Thus, increased radioactivity coming from outside this planet or from the bowels of the earth could be the cause of the spontaneous origin of new species. Should an interplanetary discharge take place between the earth and another celestial body, such as a planet, a planetoid, a trail of meteorites, or a charged cloud of gases, with possibly billions of volts of potential difference and nuclear fission or fusion, the effect would be similar to that of an explosion of many hydrogen bombs with ensuing procreation of monstrosities and growth anomalies on a large scale.

What matters is that the principle that *can* cause the origin of species exists in nature. The irony lies in the circumstance that Darwin saw in catastrophism the chief adversary of his theory of the origin of species, being led by the conviction that new species could evolve as a result of competition with accidental characteristics serving as weapons only if almost limitless time were at the disposal of that competition, with no catastrophes intervening. Now exactly the opposite is true: competition cannot cause new species to evolve. Mutations in single traits and the resulting new varieties within a species are caused by radiation hitting some gene, as did the X-rays in the experiments on the vinegar fly; it is a hit, or a collision, or a miniature catastrophe. In order for a simultaneous mutation of many characteristics to occur, with a new species as a resultant, a radiation shower of terrestrial or extraterrestrial origin must take place. Therefore we are led to the belief that evolution is a process initiated in catastrophes. Numerous catastrophes or bursts of effective radiation must have taken place in the geological past in order to change so radically the living forms on earth, as the record of fossils embedded in lava and sediment bears witness.

How would this understanding of evolution meet the facts,

and especially those facts that always appeared to be in discord with the theory of natural selection?

The fact that some organisms, like foraminifera survived all geological ages without participating in evolution, a point of perplexity in the theory of natural selection, would be explained by catastrophic evolution in which many species would be destroyed, others would be subjected to multiple mutations, and some specimens of species would escape mutations and procreate their old form.

The fact that the geological record shows a sudden emergence of many new forms at the beginning of each geological age does not require the artificial explanation that the records are always defective; the geological records truly reflect the changes in the animal and plant worlds from one period of geological time to the next. Many of the new species evolved in the wake of a global catastrophe, at the beginning of a new age, were entombed in a subsequent paroxysm of nature at the end of that age.

The fact that in many cases the intermediary links between present-day species are missing, as well as those between various species of the geological record, a vexing problem, is understandable in the light of sudden and multiple variations that gave rise to new species.

It was objected that if a new characteristic appeared in only a single animal, as the theory of natural selection claims, or even in a few animals of the same species, it would disappear in succeeding generations through interbreeding unless the new animal had been protected by isolation on secluded islands. However, in catastrophic evolution, the simultaneous mutation of many genes could produce a new species at the first fertilization; all the offspring of a litter could be affected similarly. And it is not inconceivable that in more than one creature of the same species, under similar circumstances of radiation, similar changes in the genes would occur; so in the X-ray experiments on *Drosophila*, similar mutations occurred in more than one fly.

The objection to the theory of natural selection, that the developed plan in a new species must appear suddenly or the race would expire—as in the case of the kangaroo pockets—is answerable within the framework of catastrophic evolution; however, the purposefulness of animal structures will remain a problem deserving of as much wonder as, for instance, the purposeful behaviour of leucocytes in the blood that rush to combat a noxious intruder.

The fact stressed by Agassiz that numerous earlier species of fish showed a more highly developed organism when compared with later species of fish can be explained by the destruction of earlier forms, not in the process of competition, but in upheavals against which superior structure is no defence.

The observation that healthy species of animals, like mammoths, with no sign of degeneration suddenly became extinct greatly troubled the evolutionists. This fact is unexplainable by natural selection or the principle of competition; not so by the catastrophic intervention of nature.

The fact that at several stages of the past many animals of various species and many species in toto were rather suddenly exterminated, in conflict with the idea of slow extinction in natural selection, conforms with the theory of cataclysmic evolution.

The enigmatic observation that the larger animals were particularly subject to extinction—the giant mammals that succumbed at the end of the Tertiary, and again in the Pleistocene, as earlier the dinosaurs did—is comprehensible if one thinks of the better chances smaller animals have of finding refuge from the ravages of nature.

Natural selection had its role, too, but not in procreating new species; it was a decisive factor in the survival or dying out of new forms, in the struggle for existence, not only between individuals, races, species, and orders, but also against the elements. In natural selection all those forms were weeded out that could not meet competition or the rapidly changing conditions of a world in upheaval.

The origin of new species from old could be caused by the processes that can be duplicated in laboratories—by excessive radiation or some other irritant in abnormal doses, thermal or chemical, all of which must have taken part in natural catastrophes of the past, and could have played a role in building new species, as the case of new plants in the bomb craters appears to indicate.

The theory of evolution is vindicated by catastrophic events in the earth's past; the proclaimed enemy of this theory proved to be its only ally. The real enemy of the theory of evolution is the teaching of uniformity, or the non-occurrence of any extraordinary events in the past. This teaching, called by Darwin the mainstay of the theory of evolution, almost set the theory apart from reality.

Great catastrophes of the past accompanied by electrical discharges and followed by radioactivity could have produced sudden and multiple mutations of the kind achieved today by experimenters, but on an immense scale. The past of mankind, and of the animal and plant kingdoms, too, must now be viewed in the light of the experience of Hiroshima and no longer from the portholes of the *Beagle*.[1]

[1] The ship on which Charles Darwin as a young naturalist made his voyage around the globe.

Chapter XVI

THE END

In the present book the testimony of stone and bone has been written down. We have listened to witnesses of various epochs, old and recent, of different latitudes, north and south, of various origins, from mountain peak and ocean bottoms—skeletons, and ashes, and lava. Long before the crowd of witnesses finished filing by, we knew that we would not be able to evade the conclusion that global catastrophes have shaken this world of ours. I have not included here the testimony of ancient literary sources or of folklore. Shall I be confronted with the argument that, though the geological and archaeological records speak for catastrophic occurrences in the past, the absence of human testimony contradicts this interpretation of the geological record of recent date? Is not *Worlds in Collision* a book of human evidence? And was not this testimony disputed because, first of all, of a presumed conflict with the findings of geology?

Although no references to historical inscriptions or to literary monuments of ancient times have been adduced here to show correspondence between the geological and historical records no attentive reader, not even a cursory peruser of these pages, could have read them without associating their content with that of many chapters of *Worlds in Collision*, if he had read the other book too. There the story was told of hurricanes of global magnitude, of forests burning and swept away, of dust, stones, fire, and ashes falling from the sky, of mountains melting like wax, of lava flowing from riven ground, of boiling seas, of bituminous rain, of shaking ground and destroyed cities, of humans

228

seeking refuge in caverns and fissures of the rock in the mountains, of oceans upheaved and falling on the land, of tidal waves moving toward the poles and back, of land becoming sea by submersion and the expanse of sea turning into desert, islands born and others drowned, mountain ridges levelled and others rising, of crowds of rivers seeking new beds, of sources that disappeared and others that became bitter, of great destructions in the animal kingdom, of decimated mankind, of migrations, of heavy clouds of dust covering the face of the earth for decades, of magnetic disturbances, of changed climates, of displaced cardinal points and altered latitudes, of disrupted calendars, and of sundials and water clocks that point to changed length of day, month, and year, of a new polar star.

All this was presented in *Worlds in Collision* as having taken place in two series of events, the first in the fifteenth century before the present era, or thirty-four centuries ago, the other of lesser intensity, in the eighth century, and the beginning of the seventh, twenty-seven centuries ago. Events of a similar nature and on an even more grandiose scale took place also in earlier ages. The narration of some of these events, as far as the human memory retained their recollection, is reserved for another volume, a sequel to *Worlds in Collision*.

Wherever we investigate the geological and paleontological records of this earth we find signs of catastrophes and upheavals, old and recent. Mountains sprang from plains, and other mountains were levelled; strata of the terrestrial crust were folded and pressed together and overturned and moved and put on top of other formations; igneous rock melted and flooded enormous areas of land with miles-thick sheets; the ocean bed flowed with molten rock; ashes showered down and built layers many yards thick on the ground and on the bottom of the oceans in their vast expanse; shores of ancient lakes were tilted and are no longer horizontal; seacoasts show subsidence or emergence, in some places over one thousand feet; rocks of the earth are filled with remains of life extinguished in a state of agony; sedimentary rocks are one vast graveyard, and the granite and basalt, too, have embedded in them numberless living organisms; and shells have closed valves as they do in a living state, so unexpectedly came the entombment; and vast forests were burned and washed away and covered with the waters of the seas and with sand and turned to coal; and animals were swept to the far north and thrown into heaps and were soaked by bituminous outpourings; and broken bones and torn ligaments and the skins of animals of living species and of extinct were smashed together with splintered forests into huge piles; and whales were cast out of the oceans onto mountains; and rocks from disintegrating mountain ridges were carried over vast stretches of land, from Norway to the Carpathians, and into the Harz Mountains, and into Scotland, and from Mount Blanc to the Juras, and from Labrador to the Poconos; and the Rocky Mountains moved many leagues from their place, and the Alps travelled a hundred miles northward, and the Himalayas and the Andes climbed ever higher; and the mountain lakes emptied themselves over barriers, and continents were torn by rifts, and the sea bottom by canyons; and land disappeared under the sea, and the sea pushed new islands from its bottom, and sea beds were turned into high mountains bearing sea shells, and shoals of fish were poisoned and boiled in the seas, and numberless rivers lost their channels, were dammed by lava and turned upstream, and the climate suddenly changed; tillable land and meadows turned into vast deserts. Reindeer from Lapland and polar fox and arctic bears from the snowy tundras and rhinoceroses and hippopotami from the African jungles, and lions from the desert and ostriches, and seals, were thrown into piles and covered with gravel, clay, and tuff, and the fissures of multitudes of rocks are filled with broken bones; regions where the palm grew were moved into the Arctic, and oceans steamed, and the evaporated seas condensed under clouds of dust and built mountainous covers of ice over great stretches of continents, and the ice melted on heated ground and cast icebergs into the oceans in enormous fleets; and all volcanoes erupted, and all human dwellings were shattered and burned, and animals tame and fierce and human beings with them ran for refuge to mountain caves, and mountains swallowed and entombed those that reached the refuge, and many species and genera and families of the animal kingdom were annihilated down to the very last one; and the earth and the sea and the sky again and again united their elements in one great work of destruction.

Following the trail of geology, we were led by the merciless logic of facts and figures to the conclusion that the earth was more than once a stage on which acts of a great drama took place, and no place on earth was free of its effects.

In the face of the evidence we were also compelled to concede that the most recent paroxysms of nature happened in historical times, only a few thousand years ago, when in some parts of the world civilization was already entering the Iron Age, but in other parts still lingering in the Neolithic or even in the Paleolithic, or rude stone, Age. The laminations of lakes, the salt content of those without outflow, the retreat of waterfalls, the elevation of mountains, pollen analysis, and archaeological finds, as well as the recent drop of the ocean level, all show how close to our time must have occurred the more recent paroxysms of nature.

The evidence is also overwhelming that the great global catastrophes were either accompanied or caused by shifting of the terrestrial axis or by a disturbance in the diurnal and annual motions of the earth. The shifting of the axis could not have been brought about by internal causes, as the proponents of the Ice Age theory in the nineteenth century assumed it was; it must have occurred, and repeatedly, under the impact of external forces. The state of lavas with reversed magnetization, hundreds of times more intensive than the inverted terrestrial magnetic field could impart, reveals the nature of the forces that were in action.

Thus from the geological evidence we came to the conclusion to which we had also arrived travelling the road of the historical and literary traditions of the peoples of the world—that the earth repeatedly went through cataclysmic events on a global scale, that the cause of these events was an extraterrestrial agent, and that some of these cosmic catastrophes took place only a few thousand years ago, in historical times.

Many world-wide phenomena, for each of which the cause is vainly sought, are explained by a single cause: The sudden changes of climate, transgression of the sea, vast volcanic and seismic activities, formation of ice cover, pluvial crises, emergence of mountains and their dislocation, rising and subsidence of coasts, tilting of lakes, sedimentation, fossilization, the provenience of tropical animals and plants in polar regions, conglomerates of fossiles of animals of various latitudes and habitats,

231

the extinction of species and genera, the appearance of new species, the reversal of the earth's magnetic field, and a score of other world-wide phenomena.

As important as the "world catastrophes" conclusion is, it grows in significance for almost every branch of science when, to the ensuing question, "Of old or of recent time?" the answer is given, "Of old and of recent." There were global catastrophes in prehuman times, in prehistoric times, and in historical times. We are descendants of survivors, themselves descendants of survivors. We read here a few pages from the logbook of the earth, a rock rolling in space, circling with its attendant lifeless satellite around a fire-breathing star, moving with this its primary and other revolving planets through the galaxy of the Milky Way of hundreds of millions of burning stars, and together with this entire host, through the void of the universe.

SUPPLEMENT

Worlds in Collision in the Light of Recent Finds in Archaeology, Geology, and Astronomy

AN ADDRESS BEFORE THE GRADUATE COLLEGE
FORUM OF PRINCETON UNIVERSITY ON OCTOBER 14, 1953

(revised version)

1895 and 1950: The Time was Ripe for a Heresy

Two hundred years ago, in 1773, Pierre Simon de Laplace (1748–1827), then twenty-three years old, stood before the Académie des Sciences in Paris and read a paper in which he proved the stability of the solar system: all deflections of the planets from their paths are only periodic oscillations from their mean courses; and the celestial mechanism is wound up to go on for ever.

Laplace's contemporary, Jean Baptiste Lamarck (1744–1829), set out to demonstrate in a series of works that this earth has ever been an abode of peaceful evolution, free from spasmodic disturbances, in opposition to the dominant views of his day.

These ideas of harmony or stability in the celestial and terrestrial spheres gained ground in the nineteenth century and became the foundation of scientific thought. In 1846 Leverrier, by announcing the existence of the planet Neptune, which was immediately thereafter discovered in the part of the sky indicated by him, proved the gravitational theory of Newton and the orderly universe of Laplace to be correct. However, in the same year, by detecting the anomaly in the revolution of Mercury, always accumulating in one and the same direction, he threw the first doubt on the infallibility of these very laws.

The theory of uniformity, as understood by Lamarck and Hutton and developed by Lyell, became the cornerstone of the Darwinian theory, and Darwin went so far as to say that anybody who was unconvinced by Lyell's teaching should refrain from reading the *Origin of Species*. The principle of uniformity, or the explanation of all past events in the history of the globe in terms of the processes in action in our own age, or the denial of catastrophic crises in the past, gave Darwin what he needed most for his idea of the origin of species: almost unlimited time. In order that from the struggle for existence, for competition, new forms should evolve, and that an animal like the spider with its many legs and human beings should have had a common ancestor, untold eons were necessary.

By the end of the nineteenth century the war between the theory of evolution and the theory of creation in six days, less than six thousand years ago, was concluded, with victory to the theory of evolution. The only difficulty left was, in the view of Thomas Huxley, that no really new species had appeared on the world scene since the scientific observations were made, not even in breeding experiments. The geological record, however, spoke unequivocally of the fact: in the past lived animal forms that do not live any longer, and of the forms that live in our age, many were not present in the geological past.

Laplace's theory of the origin of the solar system from a rotating nebula was replaced, by the end of the century, with a theory of a catastrophic beginning in a near-collision of the sun with another star, with debris forming the planets. But it was stressed by the authors of this new theory that the universe is orderly, and this beginning in a cataclysm was an unusually rare occurrence in the cosmos, and that the solar system is governed by the principle of stability, as annunciated by Laplace, and the earth by the law of uniformity, and the animals and plants by the law of evolution through continuity.

It appeared that, basic principles having been established, science had before it only the work of refinement in observation and in the addition of details for the perfection of knowledge; but the time of basic discoveries was over.

This was the outlook in 1895. In April of that year Fridtjof Nansen, in an attempt to discover the North Pole, reached a point less than four degrees from it. The scientific world looked upon the discovery of the North Pole as the most coveted goal still left to be attained by science.

But before Nansen, drifting from latitude 86° 14', reached his home in Norway, the scene changed. Konrad Roentgen of Würzberg discovered the X-rays or cathode rays that pass through opaque bodies. In the same year of 1895 twenty-year-old Marconi, working at the home of his father near Bologna, made the first successful experiment with wireless transmission. That year, too, Sigmund Freud published his first paper (together with Joseph Breuer), which led to a new understanding of the realm known as the subsconcious; and at the same time Pavlov made his contribution to the psychology of the reflexes.

The next year, and still before Nansen had landed on the Norwegian coast, Henri Becquerel, working on uranium, discovered the phenomenon of radioactivity. Two years later he was followed by the Curies, who discovered radium. In 1897 J. J. Thomson announced that the atom is divisible and is actually a microcosm, and he was followed by Rutherford. In 1900 Planck presented the theory of quanta, or energies dispatched in bundles or shots, and not in a continuous stream. And in the field of the origin of species, in 1900, Van Vries announced mutations in plants, observed for the first time: a process of spontaneous changes in living nature fundamentally different from the process of evolution through continuity as postulated by Darwin.

Thus in a few years, in a spectacular series of discoveries, the entire world—matter and energy and living species and the human soul—opened new horizons and everything appeared to be in incessant vibration, collisions, and transformation: the macrocosm, the microcosm, and even the subtle world of the mind, all alike.

And in 1905 Albert Einstein, then twenty-six years old, offered his understanding of the physical world, an understanding that required a new mental approach, as a testimonial that the age of basic discoveries had not ended with the victory of Darwin over the Book of Genesis.

Since then another fifty years have passed. Once more, as before the end of the nineteenth century, we are told that the fundamentals are all known; the age of basic discoveries is definitely terminated, this time for certain; and present and future generations will have to satisfy themselves with detecting details, accumulating data, and adding decimals. And though the exciting decade of 1895 to 1905 threw light on processes in matter, life and soul, processes that are certainly not inert and are marked by spontaneity and conflict, science in its various branches adjusted the new discoveries and ideas to the framework of the old great principle reigning equally in lifeless and living nature: the law of harmony and unperturbed stability. The time was ripe for heresy.

In 1950, a book, *Worlds in Collision*, created an outburst of emotions almost unprecedented in science. In the Preface to the book I wrote: "Harmony or stability in the celestial and terrestrial spheres is the point of departure of the present-day concept of the world as expressed in the celestial mechanics of Newton and the theory of evolution of Darwin. If these two men of science are sacrosanct, this book is a heresy."

I came upon the idea that traditions and legends and memories of generic origin can be treated in the same way in which we treat in psychoanalysis the early memories of a single individual. I spent ten years on this work. I found that the collective memory of mankind spoke of a series of global catastrophes that occurred in historical times. I believed that I could even identify the exact times and the very agents of the great upheavals of the more recent past. The conclusions at which I arrived compelled me to cross the frontiers into various fields of science—archaeology, geology, and astronomy. The result was a book, a prolegomenon. In its concluding pages I conceded that more problems were raised than had been solved, and I promised, always reckoning with the limitations of the individual scholar, to pursue my study into these fields too. But already the implications of the fact of great global catastrophes on the earth, one of the celestial bodies, in a time so recent, had caused my critics to assert, in the words of a Harvard astronomer, that here was the "most amazing example of a shattering of accepted concepts on record."

In the heat of the debate in the press the book was pronounced "one of the most significant books written since the invention of printing," and also "the worst book since the invention of movable characters."

Believing that an emotional atmosphere is not well suited to fruitful debate, I have entered only infrequently into the controversy. I have made short factual corrections of statements by the Astronomer Royal and by J. B. S. Haldane appearing in their reviews of my book, and I participated in a debate with your professor of astronomy, J. Q. Stewart, in the pages of *Harper's Magazine* (June 1951). I appeared before the American Philosophical Society, which at its annual meeting in April 1952 held a symposium on "Some Unorthodoxies of Modern Science," my unorthodoxy being the chief subject on the agenda. Otherwise I have kept myself out of the verbal conflict.

Now more than three and a half years have passed since the publication of the book, and I appreciate the opportunity offered me by your invitation to present a dispassionate review of recent finds in the three fields named in the title of my address.

Worlds in Collision and Recent Finds in Archaeology

In my book I described the great natural catastrophes of the second and first millennia before the present era. Prominent place is given to the description of the natural upheaval that occurred in the closing hours of the Middle Kingdom in Egypt. I synchronized this event with the Exodus, when sea, land, and sky were in uproar. The collective human memory retained an inexhaustible array of recollections of the time when the world was in conflagration, when sea engulfed land, earth trembled, celestial bodies were disturbed in their motion, and meteorites fell. My narrative is based on historical texts of many peoples around the globe, on classical literature, on epics of northern races, on sacred books of the Orient and Occident, on traditions and folklore of primitive peoples.

The question that arose was: Where is the archaeological evidence? In later chapters of my book I gave such evidence: water clocks and sundials that show a different length of the day or altered latitudes; change in the orientation of ancient temples which originally faced toward the east but do so no longer. I also closely examined in my book the calendars of the civilized peoples of antiquity, from Mexico and Peru to Greece, Iran, Israel, Egypt, Babylon, Assyria, India, and China, and the calendar reforms that were made. All this material gave strong support to the literary evidence.

Working independently of me, Professor Claude Schaeffer, whose earlier excavations at Ras-Shamra (Ugarit) caused a complete revolution in biblical exegesis, published a volume, *Stratigraphie comparée et chronologie de l'Asie Occidentale (III^eand IIe millénaires)*, printed by the Oxford University Press.[1] In this very detailed and technical work, comprising, together with tables, almost a thousand pages, Schaeffer demonstrates that on several occasions, each marking the end of an epoch, the entire ancient East was shaken and devastated. Modern annals of seismology know nothing comparable in severity and extent. The most devastating of these upheavals took place at exactly the end of the Middle Kingdom in Egypt, causing its downfall—as claimed in *Worlds in Collision* and *Ages in Chaos*.

Cities were overturned; epidemics left the dead piled in common graves; the pursuit of arts and commerce came to an abrupt end; empires ceased to exist; strata of earth, dust, and ashes yards thick covered the ruined cities. In many places the population was annihilated, in others it was decimated; settled living was replaced by nomadic existence. Climate changed.

Claude Schaeffer analyzed the archaeological finds of every place excavated from Troy at the Dardanelles over all Asia Minor, Armenia, the Caucasus, Persia, Syria, Cyprus, and Palestine to Egypt in Africa; he summarizes his extensive volume thus:

"Our inquiry has demonstrated that these repeated crises which opened and closed the principal periods of the third and second millennia were not caused by the action of man. Far from it, because compared with the vastness of these all-embracing crises and their profound effects, the exploits of conquerors . . . would appear only insignificant."

Schaeffer's work sheds a new light on the conclusions at which Sir Arthur Evans arrived after many years of archaeological work on Crete: the island was shattered in violent catastrophes that were accompanied by fire, and in these

[1] Schaeffer's book was published in 1948; my attention was drawn to it in 1951 by Dr. W. Federn. My first publication claiming natural catastrophes that overwhelmed the ancient East—the greatest of which caused the downfall of the Middle Kingdom in Egypt—was printed in January 1946 as *Theses for the Reconstruction of Ancient History*, a monograph published in the series *Scripta Academica Hierosolymitana*. It embodies the entire plan of *Ages in Chaos* in the form of a summary. As to the causes of the catastrophes, Schaeffer wrote: "Nous ne distinguons encore qu'imperfaitement les causes initiales et réelles de certaines de ces grandes crises."

upheavals the cultural and political ages of the Minoan kingdom went down, at the same time that corresponding Egyptian ages were terminated. Troy III was destroyed and covered by a fifty-foot layer of ashes when the Middle Kingdom in Egypt fell; the volcano on the island of Thera exploded with almost unimaginable fury; recent archaeological work in the Indus Valley showed, too, that about 1500 B.C., and in advance of the Arian invasion, cities with great walls were destroyed and a flourishing civilization came to a sudden end.

The synchronization of the Exodus with the end of the Middle Kingdom was also the starting point of a reconstruction of ancient history from that point on to the advent of Alexander the Great, which took the form of a two-volume work, *Ages in Chaos*, the first volume of which was published in America in the Spring of 1952. The problem of the time of the Exodus in Egyptian history had never been solved. In the Papyrus Ipuwer and the Naos of El Arish I found descriptions of a natural upheaval very similar, sometimes identical, with the description in the Book of Exodus: following plagues, when the river was blood-coloured, amid a hurricane and darkness of seven days' duration, the pharaoh and his host were drowned in a whirlpool at Pi-ha-Kiroth, the same place where the pharaoh of the Exodus was drowned. These parallels compelled me to fix an unorthodox date for the Exodus. Collating the historical texts of following generations for twelve hundred years, I could establish numerous correlations between the histories of Egypt and of Israel which could not be accidental; my reconstruction demonstrated that Egyptian history and the histories of the nations which are written in harmony with it are out of line with the historical past by about six to seven hundred years.

Thus both my works have their starting point in the recognition that the Middle Kingdom in Egypt went down in a great natural catastrophe.

The recent excavation in Jericho has confirmed the fact that the great walls of the city fell a few decades after the end of the Middle Kingdom. But at the same time in which conventional chronology places the arrival of the Israelites under Joshua in Canaan, there was no city at Jericho and no walls to fall. According to *Ages in Chaos*, however, the Israelites came to the walls of Jericho one generation after the end of the Middle Kingdom, and the enigmatic hiatus of six hundred years proves not to be real.

I expect new evidence from the Minoan scripts and the so-called Hittite pictographs. Texts in the Minoan (Linear B) script were found years ago on Crete and in Mycenae and in several other places on the Greek mainland. I believe that when the Minoan writings unearthed in Mycenae are deciphered they will be found to be Greek. I also claim that these texts are of a later date than generally believed. "No 'Dark Age' of six centuries' duration intervened in Greece between the Mycenaean age and the Ionian age of the seventh century."[2]

Before long new evidence will come from the so-called Hittite pictographic writings found in Asia Minor, Mesopotamia, and northern Syria. Since the recent discovery in Karatepe in Asia Minor of bilingual inscriptions—in ancient Hebrew and in pictographs—efforts at decipherment have entered a new stage. Today the Hittite pictographs are already in the process of being read. In my reconstruction I come to the conclusion that they are Chaldean signs, not Hittite. I also expect unequivocal evidence that these signs were used down to the last century before the present era. Owing to the confusion in the conventional chronology, the Chaldean writings of the Neo-Babylonian Empire are ascribed to early centuries and an imaginary empire.

W. F. Libby and his associates at the University of Chicago have developed the radiocarbon method of dating organic matter. Wood from under the foundation of the "Hittite" fortress of Alisar in Asia Minor turned out to be seven to eight hundred years younger than conventional chronology would allow,[3] thus giving full support to my dating. Hittite history, interwoven with Egyptian history of the New Kingdom, cannot be shortened without at the same time shortening the history of Egypt. The age of pieces of wood from the tombs of the Old and Middle Kingdoms in Egypt also proved to be in harmony with my reconstruction. However, for the decisive period—that

[2] Quoted from my *Theses for the Reconstruction of Ancient History*, published as an advance summary of *Ages in Chaos*, and referred to in the foregoing footnote.

[3] W. F. Libby, *Radiocarbon Dating* (1951), pp. 71, 102.

of the New Kingdom—no radiocarbon analysis has been made.

I suggest that some objects in the posession of museums, dating from the New Kingdom in Egypt (the dynasty of Hatshepsut, Thutmose III, Akhnaton, and Tutankhamen, and those of Ramses II and Ramses III), be subjected to the radiocarbon test.

Soon you will be able to judge as right or wrong my unqualified statement that carbon analysis of the wooden sarcophagi of Seti, Ramses II, Merneptah, and Ramses III, or of the furniture and sacred boats of Thutmose III or Tutankhamen, would yield dates five to seven hundred years younger than those assigned by adherents of the conventional chronology. Then you will know for certain whether the conventional or the revised history of the lands of the ancient East for twelve hundred years is authentic and true.[4]

In recent years, Russian archaeologists have discovered abundant remains of human culture in northeastern Siberia, in the frozen taiga where frozen bodies of mammoths are found and where nobody suspected human abodes in ages past. There was human population in northeastern Siberia in paleolithic time, in neolithic time, and in the bronze time too.

Paleolithic artifacts were found in Yakutia; rock drawings very similar to the paleolithic drawings on the rocks and in the caverns of France and Spain were found in the valley of the Lena, near the village Shishkino.

"In the neolithic age, about two to three millennia before our era, neolithic races, descendants of earlier inhabitants of

---

[4] This lecture was delivered on October 14, 1953. In November of the same year the first announcement of the decipherment of Minoan script (Linear B) was made by Michael Ventris, an English architect. Contrary to what had been thought concerning this script, it was found to be in the *Greek* language. This fact startled the scholarly world, as the texts had been erroneously referred to a time before the twelfth century. It had been generally thought that in the days of Homer, about 700 B.C., the Greeks were illiterate, and that at about that time the first attempts at writing were made in the adopted Phoenician (Hebrew) letters. The decipherment of the Minoan script forced the conslusion that a syllabic alphabet was used in Greece six hundred years before Homer. But amazement still persists, for no literary documents have come down to us from between 1300 B.C. and 700 B.C. A literate people cannot forfeit completely a well-developed literacy. As I indicated in *Ages in Chaos* and in my lecture, this period of a Dark Age of six centuries between the Mycenaean and Ionian ages results from an erroneous timetable of ancient history.

Yakutia . . . spread to the very coast of the Arctic Ocean in the north and the Koluma in the east."[5]

In *Worlds in Collision*, p. 315, I expressed my belief that human settlements would be discovered "farther to the north on the Koluma or Lena rivers flowing into the Arctic Ocean." On the lower Lena north of the confluence with Viliy, inside the polar circle, monuments are found of a characteristic culture; outstanding finds were made near the lake Yolba, not far from Jigansk.

As soon as the archaeologists started a methodic investigation of the area, in Yakutsk itself was found a workshop of an ancient metallurgist in which, at the end of the second millennium before the present era, he made bronze axes similar to the axes manufactured about that time in the Near East and in Europe.

"In the Yakutsk taiga two and a half [or three] thousand years ago, there already lived artisans in metals who were able to extract copper from ore, to melt it and pour it into forms, and to make axes, beautiful bronze tips for the spears, knives and even swords."[6]

These relics of a civilization in the taiga of northeastern Siberia imply that the climate changed there in the age of advanced man. Before the ice froze the region, voracious members of the elephant family roamed there in large herds.

Recent Finds in Geology

Archaeological evidence of continental upheavals in the second millennium having been presented in detail by Schaeffer, the evidence of geology and paleontology called for elucidation. To this I have dedicated a special work, now close to completion, and since it will be published before very long, I shall refer here only briefly to some of this material.

A little over a decade ago it was observed that the gold-digging hydraulic giants in the Fairbanks District in Alaska, sluicing out miles-long cuts, opened great hecatombs of

---

[5] A. P. Okladnikov, "Excavations in the North" in *Po Sledam Drevnikh Kultur* (Vestiges of Ancient Cultures), Gosudarstvenoye Isdatelstvo Kulturno-Prosvetitelnoy Literatury, 1951.

[6] Ibid.

animals. "Their numbers are appalling. They lie frozen in tangled masses, interspersed with uprooted trees. They seem to have been torn apart and dismembered and then consolidated under catastrophic conditions. Skin, ligament, hair, flesh, can still be seen."[1] Then human artifacts were found under the mass of torn animals and splintered trees. These artifacts do not differ much from those used only recently by the Indians of the Tanana Valley in Alaska. Mammoths, mastodons, superbison, lions, horses were found among other animals.

Since then similar finds of bones and artifacts have been unearthed all over Asia. They bring to mind the finds made long ago in the "Ivory Islands" of the Arctic Ocean above Siberia. "These islands were full of mammoth bones, and the quantity of tusks and teeth of elephants and rhinoceroses, found in the newly discovered islands of New Siberia, were perfectly amazing. . . . The soil of these isolated islands is absolutely packed full of the bones of elephants and rhinoceroses in astonishing numbers."[2] These bones are mixed with trunks of trees heaped hundreds of feet high, broken and charred.

Hippopotami, animals that live in the marshes of Africa, left their bones in abundance in England and France, and these bones are not yet fossilized. J. Prestwich, professor of geology at Oxford (1874–1888), was early struck by the finds in the fissures of the rocks in England, central and southern France, Gibraltar, and the islands of the Mediterranean.[3] Bones of animals, living and extinct, in great masses choke these fissures and caves. Some fissures are on top of high hills, and they, too, are filled with bones. The bones are broken into innumerable fragments and are still fresh; artifacts of man are found among them. Prestwich understood that some catastrophe of continental dimensions, with water playing the main role, swept over Europe in the time when the Neolithic Age started there and

---

[1] K. Macgowan, *Early Man in the New World* (1950), p. 151; cf. F. Rainey, "Archaeological Investigation in Central Alaska," *American Antiquity*, V (1940), 305; cf. F. C. Hibben, "Evidence of Early Man in Alaska," *American Antiquity*, VIII (1943), 256.

[2] D. G. Whitley, *Journal of the Philosophical Society of Great Britain*, XII (1910), 35.

[3] J. Prestwich, *Quarterly Journal of the Geological Society*, XLVIII; *Philosophical Transactions of the Royal Society of London* (1893), 1894; also see his *On certain phenomena . . .* (London, 1895).

when the Bronze Age may have been well on its way in the centres of ancient civilization.

Palms were found to have grown in northern Greenland, where now for half a year there is darkness and it is permanently cold. At some time in the remote past corals grew in Spitsbergen, and sequoia forests in Alaska; and it was early understood that the terrestrial axis must have changed its position. Airy, Lord Kelvin, George Darwin, and many others, including Schiaparelli and Simon Newcomb, participated in a long debate on the astronomical and geological possibility of a sudden change in the direction of the terrestrial axis, a debate that was erroneously thought to have been started as a consequence of *Worlds in Collision*. It was understood that such a change must have taken place unless the strange finds are to be left without explanation. The theory of drifting continents, offered as a substitute, was rejected for many reasons. Jeffreys showed that the mobile force invoked by Wegener is one hundred billion times too small to move the continents. Eddington thought that possibly only the crust, in its entirety, moved, and the axis of the core was left unchanged in direction. But the mobile force he invoked—the tidal inequalities of lunar origin—would not have moved the latitudes out of their places, the directional pull being east-west.

W. B. Wright, in his *The Quaternary Ice Age* (2nd ed., 1937), says that during geological history there occurred many changes in the position of the climatic zones on the surface of the earth which cannot be explained except by a shifting of the axis or a displacement of the pole from its present position.

But what could have brought about a change in the inclination of the terrestrial axis to the plane of the ecliptic? I discussed this question in the closing pages of *Worlds in Collision* and suggested the entrance of the earth into a strong magnetic field.

The newly developed science of paleomagnetism brought, and daily continues to bring, confirmation of the fact that lavas and igneous rocks in all parts of the world are reversely magnetized. But what is even more startling is to find that the reversely magnetized rocks are a hundred times more strongly magnetized than the earth's magnetic field could have caused them to be. H. Manley, in his review, writes:

"It may seem strange that a rock which is made magnetic by

246

the earth's field" should become so strongly magnetized "compared with the generating force. This is one of the most astonishing problems of paleomagnetism."[4]

Manley also refers to the tests made years ago by G. Folgheraiter and P. L. Mercanton on the clay of ancient Etruscan vases. They were found to have been fired when the vases were closer to the south magnetic pole; their position during the firing is known, because of the flow of the glaze; and the magnetic dip or inclination of the clay is found. Manley writes: "This implies that in the sixth century B.C. the earth's magnetic field was reversed in the Central Mediterranean area." He speaks also of a general "reversal in historical times, 2500 years ago," that must be cleared up by additional research.

Knowing from my study of ancient literary sources the proper time of exogenous disturbances in terrestrial rotation, I suspected an inaccuracy in the last sentence of an otherwise well-written article by Manley: the reversal must have occurred in the eighth century and again in the beginning of the seventh century (687 B.C.). I was gratified to find, in the original publication of Professor Mercanton, to whom I directed my inquiry, that the vases with reversed polarity date from the *eighth* century.[5]

I expect that, should the research be extended to vases dating from the end of the Middle Kingdom in Egypt (circa 3500 years ago), other periods of "unnatural" polarity would be determined in Egypt and elsewhere.

Professor R. Daly of Harvard University found that 3500 years ago all over the world the level of the oceans suddenly dropped. He thought it might be due to a sudden sinking of the crust. And in an authoritative work, *Marine Geology* (1950), Professor P. H. Kuenen of the Netherlands finds that "this recent shift is now well established" on observations in many places of the world, and he, too, assigns this catastrophic drop of the ocean level to 3500 years ago.

The recent expedition of the Oceanographic Institute at Göteborg, under H. Pettersson, which covered the Atlantic, Pacific, and Indian oceans, found, according to its leader,

---

[4] "Paleomagnetism," *Science News*, July 1949.
[5] P. L. Mercanton, in *Archives des sciences physiques et naturelles* (Quatrième Période, Tome XXIII, Geneva, 1907).

"evidences of great catastrophes that have altered the face of the earth." He speaks of "climatic catastrophes," and of "tectonic catastrophes [that] raised or lowered the ocean bottom hundreds and even thousands of feet, spreading huge tidal waves which destroyed plant and animal life on the coastal plains." At many places "a lava bed of geologically recent origin [was] covered only by a thin veneer of sediment." He discovered that the Pacific and Indian ocean beds consist "largely of volcanic ash that had settled on the bottom after great volcanic explosions." He also found a large nickel content in the clay of the ocean bottoms, and decided that this abysmal nickel must have been of meteoric origin. Consequently, he concludes, there were "very heavy showers of meteors," "The principal difficulty of this explanation is that it requires a rate of accretion of meteoric dust several hundred times greater than that which astronomers . . . . are presently prepared to admit."[6]

Professor Ewing of Columbia University carried on his investigation in the Atlantic. In 1949 he published his results, and, like Pettersson, he found that lava spread only recently on the bottom of the ocean. He also came upon signs of land deep on the bottom of the ocean and concluded: "Either the land must have sunk two to three miles, or the sea once must have been two or three miles lower than now. Either conclusion is startling."

The pollen analysis, made by various scientists, of the bottom of the North Sea, between Germany, England, Scotland, and Norway, convinced researchers that this sea in its present shape originated only very recently—in the Subboreal, the date of 1500 before the present era often being selected. At that time there occurred a *Klimasturz*. Once there had been a sea; then it was covered by debris carried from the mountains of Norway; later, in a catastrophic advance, the North Sea was formed once more. Human artifacts have been found from the time when the North Sea was land.

The investigation of the delta formation of the Bear River (on the Alaskan border), very carefully made by Hanson, showed that "at the present rate of sedimentation the delta is estimated to be only 3600 years old," A. de Lapparent, the leading French geo-

[6] H. Pettersson, "Exploring the Ocean Floor," *Scientific American*, August 1950.

logist of the beginning of the century, calculated that, since the time the Rhone glacier started to melt, less than 3000 years have passed. Modern research confirms that many of the Alpine glaciers are less than 4000 years old. Professor Flint of Yale refers to the redetermination of the age of the Upper Great Gorge of Niagara Falls and writes (1947): "The age of the Upper Great Gorge is calculated as somewhat more than four thousand years—and to obtain even this [low] figure we have to assume that the rate of recession has been constant, although we know that discharge has in fact varied greatly during post-glacial times."[7]

Sernander and others demonstrated that in 1500 B.C. and again after 800 B.C. there occurred climatic catastrophes of global dimensions. These researches, unknown to me when I wrote *Worlds in Collision*, coincide completely with my conclusions and their dating.

In both these periods the lake dwellings in Switzerland, Germany, northern Italy, and also in Scandinavia were overwhelmed by "high water catastrophes" and abandoned the first time for four centuries, the second time never to be rebuilt.

H. Gams and R. Nordhagen showed, with very extensive documentation, that at these two time dates the lakes of Europe were tilted, and many of them, like Ess-see and Federsee, were emptied of all their water. The Isartal in the Bavarian Alps was "violently torn out" and this "in very recent times"; and in the Inntal in the Tyrol the "many changes of river beds are indicative of ground movements on a great scale."[8]

H. de Terra of the Carnegie Institute and Peterson of Harvard came to the conclusion that the Himalayas, in violent upheavals, reached their present form and height in the age of man, partly even in the time of advanced man. The same conclusion is made concerning the Andes, where, too, the upheaval must have been catastrophic. In the age of man the Andes rose many thousands of feet amid volcanic activity.

In the hills of Montreal and New Hampshire and in Michigan, five and six hundred feet above sea level, bones of whales

[7] R. F. Flint, *Glacial Geology and the Pleistocene Epoch* (1947), p. 382.
[8] H. Gams and R. Nordhagen, "Postglaziale Klimaänderungen und Erdkrustenbewegingen in Mitteleuropa," *Mitteilungen der Geographischen Gesellschaft in München* (1923), pp. 13–336.

have been found. In many places on the earth—on all continents—bones of sea animals and polar land animals and
tropical animals have been found in great melees; so also in the
Cumberland Caves in Maryland and in the Choukoutien
fissure in China, and in Germany and Denmark. Hippopotami
and ostriches were found together with seals and reindeer.
Wherever we turn our interest—from the Arctic to the Antarctic
and from sunrise to sunset, in the high mountains and in the
deep seas—we find innumerable signs of great upheavals,
ancient and recent.

A circular meteoric crater (Chubb crater) was discovered in
the summer of 1950 in northern Labrador; it covers an area of
four square miles. It is much larger than the Arizona crater,
which is four fifths of a mile in diameter (two thirds of a square
mile in area); whereas the Arizona crater could accommodate
two million people in its amphitheatre, the Chubb crater could
accommodate twelve million people. It must have been created
by the impact of an asteroid. According to the published
opinion of geological authorities, the asteroid must have fallen
about four thousand years ago.

Following, or shortly preceding, the discovery of the Chubb
crater, several other large meteoric craters were discovered in
Australia, Arabia, and Mexico. The tens of thousands of oval
formations on the Atlantic coast of the United States,
especially in the Carolinas, some of them attaining a length of a
few miles each, were conclusively identified, in a monograph
by W. F. Prouty (1952), as having been caused by the fall of
large meteorites.[9] And finally, the largest crater formations,
situated in Quebec north of Sept Iles, in Canada and occupying
an area of 680 square miles, is under investigation as to its
meteoric origin by a group of Mines Department scientists led
by Dr. M. J. S. Innes.

Of the many other new developments in the field of geology,
I would stress some of the results obained by the radiocarbon
method. The time of the Ice Age is moved much closer to our
time. Instead of 25,000 years as the terminal date of the last

[9] *Bulletin of the Geological Society of America,* LXIII (1952).

glacial period, it is shown that 10,000 or 11,000 years ago the ice was still *advancing*; and even with this low dating there remain "puzzling exceptions,"[10] among them the finding of mastodons and mammoths in strata only 3500 years old. [Moreover, organic vestiges in the drift of the last glaciation have been found to be of a radiocarbon age pointing to a time 3500 years ago.[11]]

Radiocarbon analysis of oil has also shown that in the deposits of the Gulf of Mexico the age of oil is measured in thousands of years, not millions.[12] This destroys the main argument the geologists have raised against the theory of the exogenous origin of some deposits of oil.

Hydrocarbons have been identified in cometary tails by spectral analysis; also carbohydrates (edible products).[13] But here we are already outside the domain of geology and in the realm of astronomy.

Worlds in Collision and Recent Finds in Astronomy

In the years when the manuscript of *Worlds in Collision* was in the hands of The Macmillan Company of New York, accepted for publication though not yet published (1946–49), and in the years following its publication in 1950 several fundamental observations were made and explanations offered that have a clear bearing on the theory of that book.

The zodiacal light, or the glow seen in the evening sky after sunset, stretching in the path of the sun and other planets (ecliptic), the mysterious origin of which has for a long time occupied the minds of astronomers, has been explained in recent years as the reflection of the solar light from two rings of dust particles, one following the orbit of Venus, the other an orbit between Mars and Jupiter, places where, according to *Worlds in Collision*, collisions of planets and a comet took place.

[10] Frederick Johnson (chairman of the Committee on Carbon 14 for the selection of samples for analysis), "The Significance of the Dates for Archaeology and Geology," in *Radiocarbon Dating*, ed. W. F. Libby (1952), p. 97.
[11] Suess, *Science*, September 24, 1954.
[12] P. V. Smith, *Science*, October 24, 1952.
[13] N. T. Bobrovnikoff (director of Perkins Observatory), "Comets," in *Astrophysics*, ed J. A. Hynek (1951), p. 342.

The origin of asteroids, or small planets, that circle between Jupiter and Mars, some of which cross the orbit of Mars and even that of the earth, has lately been explained as the result of the explosion of a planet and more recently (1950) as the result of a collision between two planets in an early age (Kuiper). N. T. Bobrovnikoff of the Perkins Observatory offered anew his own explanation of the origin of the asteroids; they are "remnants of a gigantic prehistoric comet." F. Whipple, upon calculating the orbits of the asteroids, came to the conclusion (1950) that two collisions occurred between these bodies and a comet, once 4700 years ago and the second time 1500 years ago, or within historical times. These dates of collision in the solar system are of the same order as those offered in *Worlds in Collision*, deduced there from historical evidence and testimony. C. Tombaugh, the discoverer of Pluto, explained (1950) the dark areas and the canals of Mars as resulting from collisions of Mars with asteroids. According to *Worlds in Collision*, Mars was involved in repeated collisions with large cometary masses.

Actually, in January 1950, an explosion observed on Mars was interpreted (by Opik) as a collision with an asteroid; clouds of dust of continental dimensions rose and screened surface features of the planet.

O. Struve of Yerkes Observatory, reviewing the achievements of astronomy during 1950, wrote that "by a bizarre coincidence" in that year "a deluge of sound papers" on "collisions within the solar system" followed on the heels of *Worlds in Collision*.

There are two theories concerning the origin of lunar craters. Their size is enormous—nothing comparable is known on earth. According to one theory, these craters are the result of a collision of the moon with very large meteorites, of the size of asteroids; according to the other theory, they are volcanic formations. Both theories assume very violent events in which the celestial body closest to the earth was involved. In *Worlds in Collision* I offered the following explanation of the lunar craters, as well as of the seas of lava and the rifts on the lunar surface: During the great catastrophes, when the moon together with the terrestrial globe passed through the fabric of a great comet and again when, in the eighth century before the present era, the earth and the moon were strongly perturbed by Mars, "the moon's surface flowed with lava and bubbled into great circular formations, which rapidly cooled off in the long lunar night, unprotected by an atmosphere from the coolness of cosmic spaces. In these cosmic collisions and near contacts the surface of the moon was also marked by clefts and rifts."

If the circular formations on the moon are these bubbles which collapsed, then probably there are smaller bubbles on the moon that have not yet burst. Dr. H. Percy Wilkins, the English selenographer, actually found over forty unexploded bubbles or domes on the moon, several of which lie to the northeast of Copernicus crater; the largest of these is found within Darwin crater and is twenty miles in diameter, according to an article by F. Benario in *Vega* (1953).

I have expressed my opinion that many comets are of recent origin, and I have supported this view by reference to the frequency and luminosity of comets in the days of imperial Rome in comparison to the number of comets visible to the unaided eye in the last centuries.

This notion received vigorous confirmation in the extensive work on comets done in Soviet Russia by a leading authority on the subject, Professor S. K. Vsehsviatsky. His research reveals that periodic comets, as observed during recent decades, are losing their luminosity and their matter at a rate so rapid that fifty or sixty revolutions suffice to disintegrate a comet completely. Thus the Halley comet can hardly go back beyond 3500 years, or the year 1500 before the present era. In the last century several comets with short periods have failed to return, having apparently lost all their matter, and a few others actually fell apart before the eyes of observers.

The rapid decay of comets excludes the possibility that they have belonged to our solar system from the beginning, or from the time the planets were formed. The theory that sees in comets bodies that arrived from other solar systems has been generally abandoned. Vsehsviatsky also shows why we must reject a theory of the capture of comets from a cloud of dust and gases through which our solar system presumably passed sometime in the past. He comes to the conclusion that the comets were born in eruptions from planets, even from satellites like our moon, where circular formations indicate violent events in the past; but the main activity must have taken place on Jupiter and Saturn, the major planets, as the form of the orbits of the short-period comets suggests. This is a revival of the theory of R. Proctor, who seventy years ago ascribed the origin of the so-called Jovian family of comets—comprising the majority of comets of short periods—to eruptions from Jupiter.

The gases of Jupiter and Saturn are in violent motion despite their low temperatures; yet the velocity necessary for escape from the major planets is so great (600 kilometres per second from Jupiter) that Vsehsviatsky admits not knowing the mechanism that could in conditions presently prevailing on major planets impart this velocity to the exploded matter. Nevertheless, Vsehsviatsky insists that in the recent past conditions on these planets must have been such that this was possible, even if these conditions cannot be defined.

He emphasizes that, by casting off the exploded matter, planets must have changed their own masses and consequently their orbits. They must also have experienced recoils.

In the *Publications of the Kiew Observatory* for 1953, Vsehsviatsky says:

"The history of the planetary system was characterized, we assume, by definitely more rapid changes and more active physical processes than appeared when only gravitational interrelations in the solar system were taken into account."[1]

All this is in complete harmony with the conclusions at which I arrived in *Worlds in Collision* concerning the time (a few thousand years ago) of the birth of comets of short periods and their origin (by eruption from the planets, especially the major planets). There I also explained the forces or conditions that caused the major planets to eject the cometary masses. "The [near] collision between major planets brought about the birth of comets" (p. 355).

Now my claim, based on historical material, that the composition of the solar system was changed in historical times, is given the support of observation and calculation.

The electromagnetic nature of the universe, deduced in *Worlds in Collision* from a series of historical phenomena, is supported by another series of recent observations.

[1] S. K. Vsehsviatsky, "New Works Concerning the Origin of Comets and the Theory of Eruption," *Publications of Kiew Observatory*, No. 5. (1953), pp. 3–57.

At Evans Signal Laboratory of the United States Army Signal Corps, in Belmar, New Jersey, researchers conducting pioneer experiments on the reception of radar echoes from the moon detected noise coming from the sun. These noises point to discharges of strong potentials.

In the autumn of 1947, at the meeting of the British Association for the Advancement of Science, Sir Edward Appleton reported that radio noises coming from the sun coincide with solar flares. According to him, "a sunspot is the most powerful ultra-short-wave radio-station known, its power being much greater than a million kilowatts."[9]

In 1948 and 1949 Donald Menzel produced motion pictures of prominences or explosions of matter on the sun; they were made at the Solar Observatory in Climax, Colorado. The exploded matter rose at a very great speed to immense heights, all the time gaining in velocity, and then descended to the sun, not on a curved path as a missile would do, but by retreating on the path it had covered, comparable to a missile reversing its direction and returning to its point of departure. Moreover, the velocity of its descent was without the acceleration expected in a fall, and this too was in violation of gravitational mechanics.

It has been observed that when protuberances or surges of exploded matter on the sun run into one another, both of them recoil violently; such observation was made by McMath and Sawyer, and on another occasion by Lyot. The conclusion drawn by E. Pettit of Mount Wilson Observatory (1951) is that solar protuberances are electrically charged.

Above the protuberances "the coronal structure is often bent into the form of an arch, sometimes into several concentric arches. This is additional evidence of the electrical nature of the prominences [protuberances] and corona."[2]

In the configuration of the cometary nuclei and tails there was found "good evidence that all particles in the comet influence the motion of each other," and the configuration of the streamers in the tails of many comets "strongly indicate a mutual repulsion." Thus wrote Professor N. T. Bobrovnikoff,

[2] E. Pettit, "The Sun and Stellar Radiation," in *Astrophysics*, ed. J. A. Hynek(1951).

director of the Perkins Observatory (1951).[3] It was also calculated that the repulsion of the tails of the comets by the sun is
twenty thousand times stronger than the gravitational attraction, and the implication is that it cannot be caused by the
pressure of light, as previously thought, and that electrical
repulsion must be in action. From spectral analysis it is known
that the cometary tails do not shine merely by reflected light,
and that their light is not caused by combustion either, but most
probably is an electrical effect, comparable to the effect of a
Geissler tube.[4]

In order to explain the general magnetic field of the terrestrial globe, Dr. E. C. Bullard, of Great Britain's National Physical Laboratory, assumed (1953) electrical currents in the liquid
metal core of the earth.

The polar lights have been explained by various scientists as
electrical charges arriving from the sun. Following disturbances
on the sun there is an immediate disturbance in the ionosphere
and radio-transmission, ground currents, and the magnetic
field of the earth; there is also a second retarded but pronounced reaction about twenty-five hours later, and auroral
displays.

In 1948 Enrico Fermi explained the enigmatically high
charges of the cosmic rays as a result of the positive particles
having travelled through magnetic fields in space. In 1951
Richtmyer and Teller, following an earlier idea of Swann, explained these charges as originating in the sun: protons and
heavy nuclei could be accelerated to the enormous velocity of
cosmic-ray particles by an extended magnetic field of solar
origin. Both theories assume the existence of magnetic fields in
space. I could add to this that if the earth is a negatively charged
body the great energy with which positive charges—the cosmic
rays—rush toward the earth is not in the least enigmatic: a
negatively charged body attracts positive charges.

At Mount Wilson Observatory Harold Babcock determined
(1947) that some of the fixed stars possess general magnetic
fields of great intensity. (One of the stars was found to reverse
its polarity every nine days from plus 7000 gauss to minus 6300
gauss. This may be understood as a sign that the star is rotating,

[3] N. T. Bobrovnikoff, "Comets," ibid., pp. 327–28.
[4] H. Spencer Jones, *General Astronomy*, pp. 273–74.

turning another pole to us every nine days. The star shows no Zeeman effect in between, that is, when the observer is in the plane of the equatorial belt of the star, in the same position in which we are permanently in relation to our sun.[5])

By 1952 the Astronomer Royal, Sir Harold Spencer Jones, estimated that magnetic properties were established in more than one hundred stars, and the number of identified magnetic stars is rapidly growing.

Several years ago Dr. J. S. Hall, of the United States Naval Observatory, and Dr. W. Hiltner, of Yerkes Observatory, found that the light from certain stars is strongly polarized. It was surmised that starlight must pass through particles of magnetized interstellar dust. The question raised was why the particles of dust should all be oriented in the same direction as their magnetic axes. However, if these clouds of dust are electrically charged and in motion, the common magnetic orientation of these particles is only natural.

In June 1950 W. Baade of Palomar and L. Spitzer of Princeton offered a theory of colliding galaxies. By November of 1952 it was definitely spoken of as "a titanic collision of two giant star clusters" behind the constellation Cygnus (Swan) of the Milky Way.

The fact of the "big crash" was substantiated by strong evidence—the radio noises coming from beyond the Milky Way and sifting through it. Galaxies, each as large as the Milky Way, with numberless stars, were riding one through another, colliding and sending out a terrible S.O.S. through the universe in the form of anguished radio noises. These signals were interpreted by Baade and K. Minkowski as the reverberations of crashes on a galactic scale. After millions of years of travelling with the velocity of light, these signals reached our radio-telescopes as clearly audible noises. They left the place of the catastrophe so long ago, yet, because of the magnitude of galaxies, the collision is possibly still going on. The signals that are emitted today will reach our solar system when our sun may

---

[5] The divergent results obtained in the determination of the solar magnetic field may be due to the varying position of the earth in relation to the solar magnetic equator, which does not coincide with the solar equator or with the ecliptic. When the earth is in the plane of the solar magnetic equator, no Zeeman effect is observable and the erroneous conclusion is made that the sun possesses no general magnetic field.

have already turned into a dwarf star and our planet into clouds of dust.

Not only the fact of the collision of galaxies startled the astronomers, but even more the medium through which it became known: the colliding galaxies send electromagnetic signals, thus evidencing the electromagnetic structure of galaxies and of the very space of the universe.

By August 1953 the statement was made that another celestial host of stars was charging on a rival galaxy in the direction of the sky where we see the Crab nebula and that still another collision was going on behind the constellation Cassiopeia.

In the March 1951 issue of *RCA Review*, John H. Nelson of the engineering department of RCA Communications, Inc., announced the results of several years of careful observations on the dependence of regular radio transmission on the position of planets in the solar system. He drew graphs and wrote: "It can be readily seen from these graphs that disturbed conditions show good correlation with planetary configurations. . . . It is definitely shown that each of the six planets studied is effective in some configurations."

The press reported: "Evidence of a strange and unexplained correlation between the positions of Jupiter, Saturn and Mars in their orbits around the sun and the presence of violent electrical disturbances in the earth's upper atmosphere . . . seems to indicate [that] the planets and the sun share in a cosmic electrical-balance mechanism that extends a billion miles from the centre of our solar system. Such an electrical balance is not accounted for in current astrophysical theories."[6]

Short-wave frequencies are disturbed when Jupiter, Saturn, and Mars line up—either in a straight line or at right angles to one another. Nelson emphasized that the phenomenon "is not due to gravitational effect or tidal pulls between planets and the sun." Actually, the phenomenon indicates that the planets are electrically charged bodies.

In this connection, the older theory of a direct but unexplained relation between the revolution of Jupiter and the sunspot cycle is seen in a new light. Also, the observation of Stetson, of the Massachusetts Institute of Technology, that the moon affects radio reception—it is twice as good when the moon is

[6] *New York Times*, April 15, 1951.

under the horizon as when it is overhead—belongs in the same category as Nelson's observation of planetary influences on the ionosphere. Stetson thought this effect was caused by some radiation emanating from the moon, for a neutral moon could not produce the phenomenon.

By 1953 the strange fact was established that the solar tides in the earth's atmosphere are sixteen times more powerful than the lunar tides in the atmosphere, a fact in complete conflict with the tidal theory, according to which the action of the moon on oceanic tides is several times more powerful than that of the sun. The fifty-fold discrepancy is still without an acceptable explanation.

These are only a few of the recent discoveries that make a revision of the mechanistic concept of the universe quite mandatory.

Exactly *because* of the accuracy achieved without reckoning with forces that appear to exist, celestial mechanics, a solid work of great mathematical minds for almost three centuries, may seem even more in need of such revision. All this has little direct bearing on the story of *Worlds in Collision*, which claims only the effect to be expected if a magnetic body like the earth should come *very close* to another magnetic body. It was my scepticism concerning the infallibility of the celestial mech-anics, which assumes the celestial bodies to be electrically and magnetically sterile, that was the real cause of the emotional outburst.

Let us think of a binary or double star; both stars revolve around each other or a common centre. A half-revolution period of a few days or only hours is common. Let us assume that the stars of the binary are magnets 7000 gauss strong. It is immediately obvious that even should the electrical component of the electromagnets be disregarded these stars are not moving in a system purely mechanical.

But this is enough to render the purely mechanical celestial system fallible in respect also to single stars and equally so to the sun and its planets.

In Jupiter and its moons we have a system not unlike the solar family. The planet is cold, yet its gases are in motion. It appears probable to me that it sends out radio noises as do the sun and the stars. I suggest that this be investigated.[7]

Uranus is the only planet about which we know that, for a considerable part of its revolution, it turns one of its poles toward us. If the gases on Uranus are not in turbulent motion, but have a smooth reflecting surface, I would expect the solar light reflected from the polar regions of Uuranus to be polarized: as is well known, light reflected from the poles of a magnet is polarized.

[It is generally thought that the magnetic field of the earth does not sensitively reach the moon. But there is a way to find out whether it does or not. The moon makes daily rocking movements—librations of latitudes, some of which are explained by no theory. I suggest investigating whether these unaccounted librations are synchronized with the daily revolutions of the magnetic poles of the earth around its geographical poles.]

C. Paine-Gaposchkin of Harvard who in the last years has written many long articles against the theory of *Worlds in Collision*, in which she asserted that the celestial bodies "could not possibly possess electrostatic charges enough to produce any of the [observed] effects on motion within the solar system," now makes, in the September 1953 issue of *Scientific American*, this confession:

"Ten years ago in our hypotheses of cosmic evolution we were thinking in terms of gravitation and light pressure. . . . Tomorrow we may contemplate a galaxy that is essentially a gravitating, turbulent electromagnet."

There will be more concessions as time goes on. Our sun and its planets are not outside a galaxy; they are not unique or an exception in the plan of the universe.

I like to tell this story. Once, in the twilight hour, a visitor

[7] On April 5, 1955, at a meeting of the American Astronomical Society, Dr. Bernard F. Burke and Dr. Kenneth L. Franklin, of the Department of Terrestrial Magnetism of the Carnegie Institution, announced the unexpected discovery of strong radio noises arising from Jupiter. They found it difficult to explain the phenomenon, since no radio noises were expected from planets. The above sentence in the lecture to the Forum, predicting noises from Jupiter, was in the typescript of the draft of the lecture as deposited in January 1954 with Professor V. Bargmann of Princeton University, and also as edited by the staff of Doubleday & Company in the summer of 1954, eight months before the discovery.

came into my study, a distinguished-looking gentleman. He brought me a manuscript dealing with celestial mechanics. After a glance at some of the pages, I had the feeling that this was the work of a mathematical genius. I entered into conversation with my visitor and mentioned the name of James Clerk Maxwell. My guest asked: "Who is he?" Embarrassed, I answered: "You know, the scientist who gave a theoretical explanation of the experiments of Faraday."

"And who is Faraday?" inquired the stranger.

In growing embarrassment I said: "Of course, the man who did the pioneer work in electromagnetism."

"And what is electromagnetism?" asked the gentleman.

"What is your name?" I inquired.

He answered: "Isaac Newton."

I awoke. On my knees was an open volume: Newton's *Principia*.

This story is told to illustrate what I have said before. Would you listen to anybody discuss the mechanics of the spheres who does not know the elementary physical forces existing in nature? But this is the position adopted by astronomers who acclaim as infallible a celestial mechanics conceived in the 1660s in which electricity and magnetism play not the slightest role.

In the fields of archaeology, geology, and astronomy the last few years have brought a vast array of facts to corroborate the claim made in *Worlds in Collision* that there were physical upheavals of a global character in historical times; that these catastrophes were caused by extraterrestrial agents; and that the nature of these agents may be identified. Although I arrived at results in conflict with orthodox beliefs, yet recent years have disclosed new observations and findings, all in support and none in refutation.

What I want to impress upon you is that science today, as in the days of Newton, lies before us as a great uncharted ocean, and we have not yet sailed very far from the coast of ignorance. In the study of the human soul we have learned only a few mechanisms of behaviour as directed from the subconscious mind, but we do not know what thinking is or what memory is. And in biology we do not know what life is. The age of basic discoveries is not yet at its end, and you are not latecomers, for whom no fundamentals are left to discover. As I see so many of you today, I visualize some of you, ten or twenty or thirty years from now, as fortunate discoverers, those of you who possess inquisitive and challenging minds, the will to persist, and an urge to store knowledge. Don't be afraid to face facts, and never lose your ability to ask the questions: Why? and How? Be in this like a child.

Don't be afraid of ridicule; think of the history of all great discoveries. I quote Alfred North Whitehead:

"If you have had your attention directed to the novelties of thought in your own lifetime, you will have observed that almost all really new ideas have a certain aspect of foolishness when they are first produced."[8]

Therefore, dare.

And should even the great ones of your age try to discourage you, think of the greatest scientist of antiquity, Archimedes, who jeered at the theory of Aristarchus, twenty-five years his senior, that the earth revolves around the sun. Untruth in science may live for centuries, and you may not see yourself vindicated, but dare.

Don't persist in your idea if the facts are against it; but do persist if you see facts gathering on your side. It may be that even the strongest opposition, that of figures, will crumble before the facts. The greatest mathematician who ever walked on these shores, Simon Newcomb, proved in 1903 that a flying machine carrying a pilot is a mathematical impossibility. In the same year of 1903 the Wright brothers, without mathematics, but by a fact, proved him wrong.

In religion, the great revelations and the great authorities— the founding fathers—belong to the past, and the older the authority, the greater it is. In science, unlike religion, the great revelations lie in the future; the coming generations are the authorities; and the pupil is greater than the master, if he has the gift to see things anew.

All fruitful ideas have been conceived in the minds of the nonconformists, for whom the known was still unknown, and who often went back to begin where others passed by, sure of their way. The truth of today was the heresy of yesterday.

[8] Alfred North Whitehead, *Science and the Modern World* (New York, 1925), Chapter III.

Imagination coupled with scepticism and an ability to wonder—if you possess these, bountiful nature will hand you some of the secrets out of her inexhaustible store. The pleasure you will experience in discovering truth will repay you for your work; don't expect other compensation, because it may not come. Yet, dare.

WORLDS IN COLLISION

IMMANUEL VELIKOVSKY

'The layman, unable to deny or to accept, will find it all,
whether fantasy or fact, quite fascinating – fascinating alike
in its stupendous pictures of a world in the grip of cosmic forces,
in its parallels drawn from the annals of the ancients in many
lands, and in its vast implications . . . this extraordinary work'
Oxford Mail
75p

AGES IN CHAOS
Vol. 1

IMMANUEL VELIKOVSKY

Taking for his starting point the simultaneous physical
catastrophes described in the book of Exodus and in Egyptian
documents, Dr Velikovsky reconstructs the political and cultural
histories of the nations of the ancient world.
'His conclusions are amazing, unheard of, revolutionary, sensational.
If Dr Velikovsky is right, this volume is the greatest contribution
to the investigation of ancient times ever written'
Dr Robert H. Pfeiffer, author of *Introduction to the Old Testament*
Illustrated 75p

THE OLD STRAIGHT TRACK

ALFRED WATKINS

The Old Straight Track remains the most important source for the study of the ancient straight tracks or leys that criss-cross the British Isles – a fascinating system which was old when the Romans came to Britain.

'A remarkable book . . . it will not be long before Alfred Watkins is recognised for what he was, an honest visionary who saw beyond the bounds of his time'
John Michell, author of *The View Over Atlantis*
Illustrated £1.25

CITY OF REVELATION
On the Proportion and Symbolic Numbers of the Cosmic Temple

JOHN MICHELL

John Michell examines the numerical formula that was the essence of the sacred canon of the earliest societies – a canon that was a complete cosmology allowing an understanding of every science, a model of all reality, and thus an image of the human mind.

'John Michell's third book excels his previous two. . . . He is a genius . . . short-circuiting established channels of thought and offering a brilliant network of his own'
Time Out
50p

SEXUAL POLITICS
by Kate Millett

The seminal book in the struggle for Women's Rights. 'Supremely interesting ... Brilliantly conceived.'

60p   *New York Times*

SAVAGE MESSIAH
by H. S. Ede

The life of Henri Gaudier-Brzeska. 'A marvellous true story of poverty and genius ... superbly and simply told.'

40p *Illustrated*   *Arts Review*

A HISTORY OF MAGIC, WITCHCRAFT AND OCCULTISM
by W. B. Crow

A one volume encylopaedia of the occult, in theory and practice, from alchemy and druidism to vampires and voo-doo.

60p

WORLDS IN COLLISION
by Immanuel Velikovsky

A startling re-interpretation of the historical past based on the comparative study of ancient civilizations and literary traditions.

60p

AKHENATEN: PHARAOH OF EGYPT
by Cyril Aldred

'Enthralling ... Akhenaten, husband of Nefertiti and predecessor of Tutankhamen, remains one of the most fascinating figures in world history.'

75p *Illustrated*   *Sunday Telegraph*